T0344681

Forensic Systems Engineering

Wiley Series in Systems Engineering and Management

William Rouse, Editor

Andrew P. Sage, Founding Editor

ANDREW P. SAGE and JAMES D. PALMER
Software Systems Engineering

WILLIAM B. ROUSE
Design for Success: A Human-Centered
Approach to Designing Successful Products
and Systems

LEONARD ADELMAN
Evaluating Decision Support and Expert
System Technology

ANDREW P. SAGE
Decision Support Systems Engineering

YEFIM FASSER and DONALD BRETINER
Process Improvement in the Electronics
Industry, Second Edition

WILLIAM B. ROUSE
Strategies for Innovation

ANDREW P. SAGE
Systems Engineering

HORST TEMPELMEIER and HEINRICH KUHN
Flexible Manufacturing Systems: Decision
Support for Design and Operation

WILLIAM B. ROUSE
Catalysts for Change: Concepts and
Principles for Enabling Innovation

*UPING FANG, KEITH W. HIPEL, and D. MARC
KILGOUR*
Interactive Decision Making: The Graph
Model for Conflict Resolution

DAVID A. SCHUM
Evidential Foundations of Probabilistic
Reasoning

*JENS RASMUSSEN, ANNELISE MARK PEJTERSEN,
and LEONARD P. GOODSTEIN*
Cognitive Systems Engineering

ANDREW P. SAGE
Systems Management for Information
Technology and Software Engineering

ALPHONSE CHAPANIS
Human Factors in Systems Engineering

YACOV Y. HAIMES
Risk Modeling, Assessment, and
Management, Third Edition

DENNIS M. SUEDE
The Engineering Design of Systems: Models
and Methods, Second Edition

ANDREW P. SAGE and JAMES E. ARMSTRONG, Jr.
Introduction to Systems Engineering

WILLIAM B. ROUSE
Essential Challenges of Strategic
Management

YEFIM FASSER and DONALD BRETTNER
Management for Quality in High-
Technology Enterprises

THOMAS B. SHERIDAN
Humans and Automation: System Design
and Research Issues

ALEXANDER KOSSIAKOFF and WILLIAM N. SWEET
Systems Engineering Principles and
Practice

HAROLD R. BOOHER
Handbook of Human Systems Integration

JEFFREY T. POLLOCK and RALPH HODGSON
Adaptive Information: Improving Business
Through Semantic Interoperability, Grid
Computing, and Enterprise Integration

ALAN L. PORTER and SCOTT W. CUNNINGHAM
Tech Mining: Exploiting New Technologies
for Competitive Advantage

REX BROWN
Rational Choice and Judgment: Decision
Analysis for the Decider

WILLIAM B. ROUSE and KENNETH R. BOFF (Editors)
Organizational Simulation

HOWARD EISNER
Managing Complex Systems: Thinking
Outside the Box

Forensic Systems Engineering

Evaluating Operations by Discovery

William A. Stimson

Registered Offices
John Wiley & Sons, Inc., 111 River Street, Hoboken, NJ 07030, USA

Editorial Office
111 River Street, Hoboken, NJ 07030, USA

For details of our global editorial offices, customer services, and more information about Wiley products visit us at www.wiley.com.

Wiley also publishes its books in a variety of electronic formats and by print-on-demand. Some content that appears in standard print versions of this book may not be available in other formats.

Library of Congress Cataloging-in-Publication Data

Names: Stimson, William A., author.
Title: Forensic systems engineering : evaluating operations by discovery / William A. Stimson.
Description: Hoboken, NJ : Wiley, 2018. | Series: Wiley series in systems engineering and
    management | Includes bibliographical references and index. |
Identifiers: LCCN 2017039503 (print) | LCCN 2017042410 (ebook) |
    ISBN 9781119422761 (pdf) | ISBN 9781119422785 (epub) | ISBN 9781119422754 (hardback)
Subjects: LCSH: Failure analysis (Engineering) | System failures (Engineering) |
    Forensic sciences. | Evidence, Expert. | BISAC: TECHNOLOGY & ENGINEERING /
    Electronics / General.
Classification: LCC TA169.5 (ebook) | LCC TA169.5 .S755 2018 (print) |
    DDC 620/.00452–dc23
LC record available at https://lccn.loc.gov/2017039503

Cover Design: Wiley
Cover Image: © Digital Vision./Gettyimages

Set in 10/12pt Warnock by SPi Global, Pondicherry, India

Printed in the United States of America

10 9 8 7 6 5 4 3 2 1

To Josette,
  my love,
  my wife,
  my friend,
  my life.

Contents

Preface

> Scientific theories deal with concepts, never with reality. All theoretical results are derived from certain axioms by deductive logic. The theories are so formulated as to correspond in some useful sense to the real world whatever that may mean. However, this correspondence is approximate, and the physical justification of all theoretical conclusions is based on some form of inductive reasoning (Papoulis, 1965).

The profession of law is several thousand years old, at least. Given this history, it is quite natural that tradition would have an important role. This is especially true in English Common Law, in which precedence has a major influence on judicial decisions. During the past 100 years or so, product liability has developed as the basis of tort law when there is a question of harm caused by a product or service, and thus enjoys the influence of tradition. During much of this time, production volume was relatively low, claims were low in proportion, and over the years, litigation involving product liability became relatively straightforward.

Today, production volume can be massive—hundreds of thousands of units produced and sold annually, with claims increasing in proportion. The result has been class action suits and large volume manufacturing suits, all continuing to be prosecuted by product liability, one claim per unit. From an engineering point of view, this process is inefficient and even ineffective. As seen by engineers, a far more effective mechanism for litigation would be *process liability*.

The concept of process liability was first defined by attorney Leonard Miller (5 New Eng. L. Rev. 163, 1970) in his article, "Air pollution control: An introduction to process liability and other private actions." Being unschooled in law, I do not know the present status of this idea in legal circles, but it is certainly helpful in forensic analysis and in systems engineering. In this book, process liability is shown to be a direct result of systems engineering procedures and methodologies applied to business operations.

Engineers have long recognized the strong correlation of process to product and many mathematical models are commonly used that can validate this cause and effect relationship. Process liability provides a needed legal basis in forensic application. *Forensic Systems Engineering* offers a complete approach to the investigation of large volume operations by uniting the concept of process liability to systems engineering.

Organization of the Book

The purpose of forensic systems engineering is to identify dysfunctional processes and to determine root causes of process failure, and further, to assist the court in determining whether harm or a breach of contract has occurred. Chapters 1 through 6 describe the role of management in operations. Chapters 7 through 11 unite liability to the essential characteristics of processes used in these operations. Chapter 12 is a fictional case study of a manufacturer, albeit based on actual events. The narration of the study is similar to the narrative technique used in many graduate schools of business.

Chapters 13 through 15 offer formal mathematical models, widely accepted in systems engineering, to demonstrate the correlation of process to product in terms of the risk of liability. Chapter 16 delves into the most troubling area found in my years as a consultant and expert witness in the litigation of business operations—the verification and validation of processes. Chapter 17 discusses the difficulty of supplier control in the age of offshore outsourcing and supply chain management. Chapter 18 addresses an unavoidable aspect of process evaluation via discovery, the effect of sampling. Finally, Chapter 19 discusses the process of identifying nonconformities in discovery and how to assess them.

Appendices A through F provide certain basic information to the reader in those subjects that are essential to forensic systems engineering and analysis. Appendices A and B are detailed accounts of engineering issues that occur more frequently in contract litigation than others. Appendix A concerns design and development; Appendix B concerns product reliability and should be considered by the reader as a prerequisite for Chapter 10.

Appendices C through F address the statistical nature of production and service processes and the fact that a forensic audit of discovery is effectively a sampling process. Therefore, the procedures of sampling and of statistics apply. These appendices, too, should be perused before Chapter 18, and they would be helpful in understanding Chapters 13 through 16. These latter chapters introduce the subject of risk, which is a probability, and employ various mathematical models of random variables.

Definitions and Terms of Art

One of the things that I admire about the profession of law is that when a specific idea requires a unique definition, it is expressed in Latin. Examples abound: *nolo contendere*, *habeas corpus*, *qui tam*, and so on. The terminology is effective because it is constant over time and does not compete with the common language. Unfortunately, engineering lacks this insight. When engineers want to express a specific idea, they borrow terms from the common language even though the engineering definition may have little to do with common understanding. One example will suffice. A system is called *controllable* if it can be taken from an initial state to any other state in finite time. I have witnessed a meeting at NASA aborted because someone used the word "controllable" in its general meaning, thereby confusing the conversation.

In addition, even terms within engineering context vary in their meaning, depending upon the audience. The meaning of terms such as production, operations, process, and system may differ from one group to another in the business and technical community. Therefore, to prevent confusion I have provided the definition of certain technical terms as they are intended in this book.

Discovery

Discovery is a pretrial procedure in a lawsuit in which each party in litigation, by court order, may obtain evidence from the other party by means of discovery devices such as documents, interrogatories, admissions, and depositions. The term "discovery" hence refers to the body of evidence available to each party in their pursuit of justice.

Production, Service, and Operations

For brevity, in this book the phrase "production or service" is called "operations." On occasion, I may use "production" in lieu of "operations," but only if the context is manufacturing. Or I may use the term "product" when speaking of operations in accordance with common usage. For example, I may speak of product quality or product reliability even though I implicitly include service, and ask the reader to bear in mind that service also has the traits of quality and reliability that apply to production. From a systems viewpoint, there is little or no difference between production and service. For this reason, for additional brevity I may use the term "unit" in place of the phrase "product or service." For example, I might say 10 units proved to be nonconforming to requirements. These units could be 10 jet engine fan blades or they could be 10 billing accounts, depending on the context of the discussion.

Management System

The classical role of management is described in five functions: plans, organization, coordination, decision, and control (Laudon & Laudon, 1991). It is reasonable to assume that a systematic approach to these activities will optimize the effectiveness and efficiency of their results. Such an approach is called a management system. The overall system includes structures for self-correction and for improving performance. The functions become subsystems of the management system, whose role is to achieve a synergistic direction to corporate goals.

With a system of management, operations can be conducted in an orderly fashion such that responsibility, authority, and accountability may be assigned with documented procedures and traceable results. The documentation and traceability do more than provide a basis from which risk assessment and methods of improvement can be made. They also provide forensic evidence if litigation arises. The evidence may support the defense or the plaintiff, depending on its nature.

The effectiveness of management will be a result of this system. Critics claim that too strict an adherence to formal procedures will stifle innovation. On the other hand, no system at all invites fire drill modes and chaos. Forensic systems engineering will measure the effectiveness of a management system in litigation by its conformity to contract requirements. The justification for this strategy is developed throughout this book.

Performance Standard

A management system has both form and substance. The form might derive from a standard of management. In this book, frequent reference is made to standards of management whose objective is the effective performance of operations in assuring the quality of the product or service rendered. Systematic operation is essential to effectiveness and can be enhanced by management standards. Such a standard is often called a quality management system (QMS) because its purpose is to improve the quality of whatever is being produced or served. For example, ISO 9001 is such a standard.

It is not unusual that in describing a document, the words management, performance, standard, and quality all occur in the same paragraph. To minimize this repetition, I may refer to such a document according to the characteristic being discussed and call it a standard of quality management, a standard of performance, or a standard of operations. In all cases, I am talking about the same thing—the effective management of business operations.

In short, I equate a standard of performance to a standard of quality management. This convention may be controversial because "quality" has, in industry, a nebulous definition. Many a company sharply distinguishes between its operations function and its quality function. Yet, assuming that a process is causal, then quality either refers to the goodness of operations or it has little meaning. (Some might argue whether a process is causal, but engineers do not and this book goes to great lengths to demonstrate the causal relation between process and product.) I regard ISO 9001 has a parsimonious set of good business practices and therefore an excellent performance standard, recognizable as such in a court of law.

A system and a standard for that system have a straightforward relationship—that of form and substance. We might say that *form* is a model of something; *substance* is the reality of it. Philosophically, the entity may or may not have physical substance. A violin can be substantive, but the music played on it may also be substantive. Relative to standard and system, the former provides the form and the system provides the substance. Both are deemed necessary to effective performance and the forensic evidence of nonconformity in either can lead to product or process liability.

A forensic investigation is akin to an audit in that it compares the descriptive system to the normative—what it is to what it should be. An effective examination of evidence will reveal what the system is doing; what it should be doing requires a relevant standard. In forensic analysis of operational systems, any recognized performance management standard can serve this role. By "recognized," I mean a standard that is recognized within the appropriate industry and by the law. Chapter 2 provides a list of several well-recognized performance standards that would carry weight in a court of law. All of them are very good in enhancing the effectiveness of operations, but not all of them are general enough to cover both strategic and tactical activities. A standard is needed for the purpose of explaining forensic systems engineering and ISO 9001 (2015) is selected as the model standard of this book because of its international authority.

I must admit that the selection of ISO 9001 as the standard of performance for this book is taken with some unease. This standard is rewritten every few years, not in its fundamentals but in its format. A good practice in, say, Clause 3 of one year may appear in Clause 5 in another year and perhaps even under another name or with a slightly different description. I beg the reader to understand that in this book, a reference to an ISO 9001 control or to its information refers to an accepted universal principle or action and not to a particular clause, paragraph, or annual version. For forensic purposes, any reference to ISO 9001 can be defended in court, although tracking down the itemized source may take some ingenuity.

Process Liability

The notion of process liability as it applies to operations is discussed in considerable detail in Chapter 6, but the subject is crucial to forensic systems engineering and appears often in various chapters of this book as it is applied to different situations. At this point, I shall not present the argument for process liability but simply introduce its genesis.

In his paper cited earlier, attorney Leonard A. Miller introduced the concept of process liability and traced legal precedents that justified its use. With permission of Mr. Miller and of the New England Law Review, several paragraphs are extracted from his paper and inserted in this book because of their pertinence to forensic investigation. Although referring to pollution control, his arguments for process liability are logically and clearly applicable to nonconforming or dysfunctional processes, as explained in Chapter 6.

Controllability, Reachability, and Observability

Formally, a system is *controllable* if it can be taken from any initial state in its state space to its zero state in finite time. A system is *reachable* if it can be taken from the zero state to any other state in finite time (Siljak, 1969). Over the years, the need to distinguish between system controllability and reachability has lessened and the latter has largely disappeared, simply by making a minor change in the definition of controllability to include the property of reachability. This explains the earlier definition I used in talking about the engineering use of common words: Engineers today say that a system is *controllable* if it can be taken from any initial state to any other state in finite time.

A system is completely observable if all its dynamic modes of motion can be ascertained from measurements of the available outputs (Siljak). Observability is no small issue in forensics because of its relation to verification and validation, which obviously require the property of observability. From an engineering point of view, inadequate processes of verification and validation render a system unobservable and are major nonconformities in management.

Process and System

The terms *system* and its kin, *process*, have no standard meaning in business and industry. Historically, they have carried different connotations and still do. For example, the international management standard, ISO 9000 (2005), distinguishes between them, defining a *process* as a set of interrelated or interacting activities which transforms inputs into outputs, and defining a *system* somewhat differently, omitting the dynamic sense assigned to a process.

In systems theory, they are regarded as the same thing. R.E. Kalman et al. (1969) defined a system as a mathematical abstraction—a dynamical process consisting of a set of admissible inputs, a set of single-valued outputs, all possible states, and a state transition function. Since a system is a dynamical process in systems theory and a process is dynamical by definition of ISO 9000, the terms are considered equivalent in this book. I may use "process" and "system" where they are traditionally used, but I ask the reader to bear in mind that they behave the same way. The elements that compose a process or system may be called a subprocess or subsystem.

Product and Process Quality

Over the years there have been many definitions of "quality" when referring to a product, but the international definition used in this book is provided by ISO

9000 (2005): *quality is the degree to which a set of inherent characteristics of a product or service fulfils requirements.* Conformity is the fulfillment of a requirement; nonconformity is the nonfulfillment of a requirement. The requirements may denote those of a unit, customer, or the QMS. These definitions are also used in this book because they are good ones, implying how one might measure quality.

However, from a systems view, the definition of quality is necessary but not sufficient, as it infers nothing about the system that provides the product or service. One of the major objectives of this book is to demonstrate a causal relation between the conformance of a process and the conformance of its output. Any definition of quality should accommodate this relationship. Therefore, in Chapter 5, I offer an additional measure of "quality": it refers to the effectiveness of productive and service processes in assuring that products and services meet customer requirements.

Acceptable Quality and Acceptable Performance

In the context of product and process, manufacturing uses two similar terms. Recognizing that no process is perfect, industry employs the metric, *acceptable performance level* (APL), defined as the lowest acceptable performance level of a function being audited (Mills, 1989). However, the term is not used in reference to a performance objective, but it is used simply to determine a sample size.

Similarly, recognizing that no sampling plan is perfect, industry employs the metric, *acceptable quality level* (AQL), defined as the largest percent defective acceptable in a given lot (Grant & Leavenworth, 1988). From the standpoint of auditing controls, the two criteria are essentially identical. Therefore, in this book the term, acceptable performance level, is preferred when referring to either concept because it has a greater sense of systems activity, suggesting both a dynamism and a broad perspective, in keeping with systems thinking.

Effectiveness and Efficiency

In litigation, it is critical that the meaning of a term be clear and unambiguous. I generally follow the definitions of ISO 9000 (2005). *Effectiveness* is the extent to which planned activities are realized and planned results are achieved. *Efficiency* is the relationship between the results achieved and the resources used. Briefly, then, effectiveness is a measure of how good the process is; efficiency is a measure of the cost to obtain that goodness.

Compliance and Conformance

Because financial auditing is subject to legal review, its procedures are well developed and formal. They are acknowledged and respected in courts of law.

Forensic systems engineering is fundamentally an audit of evidence in discovery and as such the analysis should be conducted in a manner acceptable in court. Therefore, I often refer to the techniques of financial auditing in this book and use some of its terms, although they may differ somewhat from their meaning in business operations. *Compliance* is one such term.

A financial auditor audits financial reports for compliance to legal requirements. This corresponds closely with the definition of compliance used in manufacturing or service operations: Compliance is the affirmative indication or judgment that the *performer* of a product or service has met the requirements of the relevant contract, specifications, or regulation (Russell, 1997). In contrast, the same source defines conformance as the affirmative indication or judgment that *a product or service* has met the requirements of the relevant contract, specifications, or regulation.

Because of the kinship of process and product in liability, I continue with this kinship in performance and usually speak of the conformance of a control rather than of its compliance. This assignment can get complicated if the control is nonconforming because of misfeasance, which suggests that the control is noncompliant also. At the end of the day, the wording to be used in litigation will be determined by attorneys and not by forensic analysts or engineers.

Framework and Model

The word *framework* has several meanings but the one used quite often in business is that of a basic structure underlying a system, concept, or text. You see the word several times in Table 2.1, used in the titles of recognized performance standards. Engineers, however, tend to use the word *model* possibly because any concept is modeled mathematically before it is physically constructed. Although the two words come from different spheres, they meet in this book and are used interchangeably. Both refer to an organization or structure of elements assembled to affect a purpose. In short, they depict systems.

Sidestepped Definitions

There are several subjects of common occurrence in most civil litigation whose use cannot be avoided, but whose definitions I choose to leave unsaid. *Strict liability* and *due diligence* are used in this book in the sense that I understand them. However, I am unschooled in law and prefer that readers look up the meaning of the terms on their own.

Another such term is *standard of care*. This issue is critical to any critique of management performance and I use it often. Standard of care refers to the watchfulness, attention, caution, and prudence that a reasonable person in the circumstances would exercise. Failure to meet the standard is negligence, and any damages resulting there from may be claimed in a lawsuit by the injured party. The problem is that the "standard" is often a subjective issue upon which reasonable people can differ. I believe that in any specific litigation, standard of care will be decided by the court, so the very general description just given will do for this book.

Redundancy

The reader will find a certain amount of repetition of information in this book, and deliberately so. First, I believe that redundancy is a good teaching tool. Secondly, some important properties, understood in one context, may also be applicable in other contexts. For example, ISO 9001 is regarded internationally as a set of good business practices and this role is important from a number of points of view, each view expressed in a different chapter: Chapter 4, Chapter 5, and Chapter 8. Also, internal controls are defined redundantly: Chapter 5, Chapter 11, Chapter 14, and Chapter 15. As an additional example, a comparison of the terms *durability* and *reliability* is made both in Chapter 2 and in Appendix B because the difference is very important and not all readers will read the appendix.

References

ANSI/ISO/ASQ (2005). *ANSI/ISO/ASQ Q9000-2005: Quality Management Systems—Fundamentals and Vocabulary*. Milwaukee, WI: American National Standards Institute and the American Society for Quality.

ANSI/ISO/ASQ (2015). *ANSI/ISO/ASQ Q9001-2015: American National Standard: Quality Management System Requirements*. Milwaukee, WI: American National Standards Institute and the American Society for Quality.

Grant, E. L. and Leavenworth, R. S. (1988). *Statistical Quality Control*. New York: McGraw-Hill, p. 452.

Kalman, R. E., Falb, P. L., and Arbib, M. A. (1969). *Topics in Mathematical System Theory*. New York: McGraw-Hill, p.74.

Laudon, K. C. and Laudon, J. P. (1991). *Management Information Systems: A Contemporary Perspective*. New York: Macmillan, p. 145

Miller, L. A. (1970). "Air Pollution Control: An Introduction to Process Liability and other Private Actions." *New England Law Review*, vol. 5, pp. 163–172.

Mills, C. A. (1989). *The Quality Audit*. New York: McGraw-Hill, p. 172.

Papoulis, A., (1965). *Probability, Random Variables and Stochastic Properties*. New York: McGraw-Hill.

Russell, J. P., ed. (1997). *The Quality Audit Handbook*. Milwaukee, WI: ASQ Quality Press, p. 12.

Siljak, D. D. (1969). *Nonlinear Systems: Parameter Analysis and Design*. New York: John Wiley & Sons, Inc., pp. 445–446.

1

What Is Forensic Systems Engineering?

Forensic systems engineering can be defined as the preparation of systems engineering data for trial. This snapshot raises more questions than it answers because neither "systems" nor "systems engineering" enjoy general agreement on what they mean. *Forensics* is well defined and refers to the application of scientific knowledge to legal problems. Conversely, *systems engineering* has no generally accepted definition and each discipline holds its own parochial notion of it. When I entered the systems engineering program at the University of Virginia, only one other university in the United States offered a degree in the field.

In the computer world the term refers to the design and implementation of hardware and software assemblies. In the Department of Defense, it refers to large human machine structures. Systems theory itself is substance neutral and can be applied with equal vigor to computers, machines of war, video assemblies, banks, institutions, dams, churches, governments, ports, and logistics—any dynamic activity with a determined goal. So let us begin with the meaning of a *system*.

1.1 Systems and Systems Engineering

The Kalman et al. (1969) definition of a system, shown in the Preface, is repeated here for facility: a dynamical process consisting of a set of admissible inputs, a set of single-valued outputs, all possible states, and a state transition function. Two of these criteria are particularly significant in forensics. Admissible inputs are essential to the proper operation of a system and will be

Forensic Systems Engineering: Evaluating Operations by Discovery,
First Edition. William A. Stimson.
© 2018 John Wiley & Sons, Inc. Published 2018 by John Wiley & Sons, Inc.

important characteristics in forensic investigation. An admissible input is one for which the system was designed. The output of a system subject to inadmissible inputs is not predictable and may be nonconforming to requirements.

In practical operation, the second Kalman criterion of a state transition function refers to that mechanism by which the system changes its state. This mechanism must be controllable and observable. An implementation or change to a system that frustrates these qualifications implies a nonconforming system. In this book, then, I define a "system" as a dynamical process consisting of a set of integrated elements, admissible inputs, and controllable states that act in synergy to achieve the system purpose and goal.

At the University of Virginia's School of Engineering and Applied Science, Professor John E. Gibson (2007) defined systems engineering as "operations research plus a policy component," the latter adding the question of "why?" to the engineering "how?" Operations research (OR) concerns the conduct and coordination of operations or activities within an organization (Hillier & Lieberman, 1990). The type or nature of the organization is immaterial and operations research can also be applied to business, industry, the military, civil government, hospitals, and so on. As a forensic examination will compare what is to what should be, then contracts too, may be of concern.

In this book, then, we define *systems engineering* as the creative application of mathematics and science in the design, development, and analysis of processes, operations, and policies to achieve system objectives.

1.2   Forensic Systems Engineering

Forensic analysis is the application of scientific principles to the investigation of materials, products, structures, or components that fail or do not function as intended. Situations are investigated after the fact to establish what occurred based on collected evidence. Forensic analysis also involves testimony concerning these investigations before a court of law or other judicial forum (Webster, 2008).

If a failing material, product, structure, or component is a subset of a system, then the cause of failure may be the system itself, and if so, failure may be systemic. Therefore, the system itself is properly within the purview of forensic investigation. This idea is particularly important when the failed unit is mass produced and sold widely. An investigation of a failure in Baltimore will say nothing about a similar failure in Miami or elsewhere; there may be systemic failure and only a system level investigation will reveal it. If so, then the productive system, too, requires forensic investigation because only a system-level approach can determine the root cause.

Therefore, I define *forensic systems engineering* as the investigation of systems or processes that have failed to achieve their intended purpose, thereby causing personal injury, damage to property, or failure to achieve contracted requirements. The basis of the investigation will be the fit of the system in litigation to contract requirements according to systems engineering criteria. Established systems theory and system analytical tools are used extensively in the investigation. Such tools include probability models, linear and nonlinear programming, queuing theory, Monte Carlo simulation, decision theory, mathematical systems theory, and statistical methods. The purpose of forensic systems analysis is to identify dysfunctional processes, to determine root causes of process failure, and to assist the court in determining whether harm or a breach of contract has occurred.

Forensic systems engineering includes all of the components of engineering: design, development, operations and analysis, and synthesis. Forensic systems analysts will aid counsel in designing and developing case strategy, based on preliminary overview of findings. In this pursuit, they will use formal scientific methods in analysis of evidence.

All products are manufactured and all services organized through business processes that are integrated so as to contribute to a symbiotic or synergistic goal. In this way the processes compose an operational *system* and systems theory applies. The consequences of system failure can be dealt with by a new legal concept called *process liability.*

For ease of reference, I repeat from the Preface that in this book the term *operations* refers to both production and service. Operations can be managed in an orderly fashion with a systematic approach using well-recognized good business practices, which we shall call a management system. A company with a formal management system has the best opportunity to consistently meet or exceed customer expectations in the goodness of its products and services.

A management system should be synergistic—the parts work together effectively to achieve the system goal. All the productive and supportive activities in the company are integrated and coordinated to achieve corporate goals. All productive processes should be organized in the natural flow of things and supported with necessary resources. This type of structure is called the process approach and is suitable to the newer standards of performance. Also, the performance of the management system is continually measured for effectiveness and efficiency, with structures for improving performance.

The forensic systems analyst should understand the relationship between system and standard. Simply put, a standard is a model—pure form. It does nothing, but it enables things to be done well. Performance standards offer consistency, efficiency, and adequacy. The system is the implementation of the standard and provides the substance. Properly implemented, the system will work well and get things done to the satisfaction of the customer, performer, and shareholder.

An investigation is akin to an audit in that it compares the descriptive system to the normative—what it is to what it should be. As discussed in the Preface, the ISO 9000 (ANSI/ISO/ASQ, 2005) Quality Management System is chosen as the standard model of this book to be used in such comparisons. However, in the absence of a contract requirement for a specific standard, any equally capable standard may do, even a locally developed one. The issue in litigation is whether the performer is prudent in standards of care and due diligence.

In 1970, attorney Leonard A. Miller presented a paper in the *New England Law Review* that introduced the concept of process liability and traced legal precedents that justified its use. With permission of Mr. Miller, several paragraphs are extracted from his paper and inserted into Chapter 6 because of their pertinence to forensic investigation. Although referring to pollution control, his arguments for process liability clearly apply as a consequence of nonconforming or dysfunctional processes.

Businesses differ in how they refer to their core entities: systems, processes, cost centers, activities, and business units, to name a few. In this book, these terms may be used interchangeably to accommodate the various backgrounds of readers. But whichever terms are used, their dynamic property does not change. Whatever it is called, a system is designed to use states and feedback loops to change admissible inputs into specified outputs. It consists of the resources, inputs, outputs, and feedback mechanisms necessary to make the process work correctly and consistently. The conditions required by every system are (i) the input must be admissible—appropriate to the system design; (ii) the states of the system are established by proper set up; (iii) the feedback loop provides the capability to compare what is to what should be; (iv) and the outputs are in agreement with system objectives. In litigation, each of these conditions can be challenged and may be the focus of forensic investigation.

In sum, a performer offers to provide a product or service to a customer. A contract is drawn up listing customer requirements and the intended use of the product or service. There may also be specifications on the performance itself, such as constraints of cost and time, or the requirement to perform in a certain way, say in accordance with given industrial or international standards. In the event of customer disappointment, a forensic investigation may be called for in which it becomes apparent that a process or operation may be a contributing factor in an adverse outcome. Both plaintiff and defense attorneys may well consider forensic system engineering in their strategies.

References

ANSI/ISO/ASQ (2005). *ANSI/ISO/ASQ Q9000-2005: Quality Management Systems—Fundamentals and Vocabulary*. Milwaukee, WI: American National Standards Institute and the American Society for Quality.

Gibson, J. E., Scherer, W. T., and Gibson, W. F. (2007). *How To Do Systems Analysis*. Hoboken, NJ: John Wiley & Sons, Inc.

Hillier, F. S. and Lieberman, G. J. (1990). *Introduction to Operations Research.*
New York: McGraw-Hill, p. 5.

Kalman, R. E., Falb, P. L., and Arbib, M. A. (1969). *Topics in Mathematical System Theory.* New York: McGraw-Hill, p. 74.

Miller, L. A. (1970). "Air Pollution Control: An Introduction to Process Liability and Other Private Actions." *New England Law Review*, vol. 5, pp. 163–172.

Webster, J. G., contributor (2008). "Expert Witness and Litigation Consulting." *Career Development in Bioengineering and Biotechnology*, edited by Madhavan, G., Oakley, B., and Kun, L. New York: Springer, p. 258.

2

Contracts, Specifications, and Standards

I repeat an earlier statement that every investigation is essentially an audit—a comparison of what is to what should be. A forensic investigator will examine the evidence with that perspective in mind. From this viewpoint, any harm, injury, or breach resulting from a failed unit or from a contracted performance is an effect. The root cause will be determined by investigating the performance according to accepted industrial practice. For example, a metallurgist may examine a failed jet vane for metal fatigue. A systems analyst may examine the nonconformance of a process. In every case the forensic systems analyst must have an understanding of the applicable references: contracts, specifications, and standards.

2.1 General

In law, a contract is a formal agreement between parties to enter into reciprocal obligations. It is not necessary that a contract be in writing; verbal contracts are equally enforceable in law. However, in this book we are concerned with written contracts, and in particular, with contracts of performance. One party agrees to pay another party to do something, usually in a certain way and within a specified time. The first party may be called the customer; the second the performer, provider, or in this book, the company.

Forensic Systems Engineering: Evaluating Operations by Discovery,
First Edition. William A. Stimson.
© 2018 John Wiley & Sons, Inc. Published 2018 by John Wiley & Sons, Inc.

Necessarily then, conditions are imposed upon the performance. These conditions are called specifications because they specify what must be done. Specifications are not always expressed in numbers, but very often it is practical to do so. Numbers help to demonstrate to the performer exactly what must be done and to the customer that the thing done is exactly what was wanted. Numbers also help to achieve repeatability.

As an example, a family might hire a tutor to educate their children. A curriculum is agreed upon, with a schedule and the education begins. This type of contract can be executed with no numbers assigned at all. However, if numerical grades are assigned to the test scores, then the family receives a measure of the effectiveness of the education. Similarly, a customer might want a blue dress. No number is involved. But a specific blue can be identified with a number, perhaps a wavelength, which then enables the customer and performer to agree on expectations and also enables reproduction of the dress.

Sometimes a number must be specified. Suppose that a customer wants a fast car. A fast car cannot be built. The performer must have an idea of what the customer means by "fast," and that requirement is best identified with a number. This example demonstrates a condition that occurs more often than not. A customer wants something and quite often expresses this desire qualitatively. For example, the customer may want fresh vegetables, a durable sofa, an efficient washing machine, or an impressive business suit. The performer can provide or manufacture all of these things and to many customers. Inevitably though, for optimum customer satisfaction and for repeatability, all these things must somehow be expressed quantitatively. Freshness of vegetables is often measured in days from picking; durability of a product is often measured in mean time to failure. Efficiency of an operation can be measured in cost per use. A metric for impressiveness presents a challenge, but the cost of the item as indicated by the brand name has been shown to be effective.

In negotiating the contract, the customer is concerned with how well the job will be done. It is cause for concern if the performer has never done this job before. Usually, the customer will want a performer of some experience. This means that the performer has done the job repeatedly and has developed a set of procedures to ensure the quality and cost of the task. Repetition implies that a standard way of doing business has been developed. The standard may be in-house, that is, unique to that performer, or it may be a set of general good business practices used by many performers engaged in similar activities.

Good business practices have been codified into standard procedures by a large number of industries and institutions in order to improve the capability and professionalism of the industry and to better achieve the expectations of customers. Simply put, it is good business to use good business practices. These practices apply to how things are made and how they are performed. Standards of how things are made are called product standards. There may be legal requirements imposed upon product standards, especially if the product is a drug or medicine. Standards of how things are done are called performance standards. This book is primarily concerned with a certain kind of performance—management standards.

Some standards are simply common sense, although they may vary from country to country. For example, in the United States, the contacts on electrical appliances have two flat pins, usually polarized. In Europe the contacts are round. Some societies adopt standards that meet the requirements of their customers but may not meet others. In recent years, the trend is to international standards. For example, desktop computers are often produced that can perform anywhere that meets their power input requirements. Telephone systems, too, are designed according to internationally agreed standards to enable worldwide conversation. Thus, contracts, specifications, and standards are inseparably entwined. Sometimes both customers and performers make the mistake of treating these issues as separate entities. This mistake is grave and can lead to customer disappointment. The contract must record exactly what the customer wants and what the performer can deliver and must ensure an agreement between them. The specifications must be correct translations of the customer's requirements, which are not easy to do because quite often the customer requirements are qualitative and the specifications quantitative. The numbers may mean little to the customer, which complicates customer review and approval. Therefore, performance must be done in an acceptable way, in accordance with customer requirements, industry standards, and government regulations.

2.2 The Contract

2.2.1 Considerations

The contract must include all the obligations of the signatory parties, in unambiguous terms. You get what is in the contract and nothing more. For example, in the 1980s, the US Navy became concerned about the quality of ship repair in private shipyards, many of whom with little experience in repairing modern fighting ships with digital systems. To resolve this problem, the Navy introduced a standard of management to be invoked in the repair contracts of shipboard combat systems (United States Naval Sea Systems Command, 1989). However, the Navy's low-bid process ensured that the job scope was always underestimated and that the number of persons needed to supervise the tasks was often understated. In frustration, the team responsible for the standard, myself included, rewrote the standard to require that at least three managers would be assigned to a ship combat system job. We got what we had asked for—exactly three managers, no matter the size of the job. Whether the project was for $8 million or $80 million, only three persons were assigned to manage the work.

In sum, written contracts must include unambiguously all applicable specifications and standards needed to accomplish the contract to the satisfaction of the customer. Standards provide the performance requirements and the guidelines of good business practices. Properly written specifications exactly describe the customer requirements.

Customer requirements are those needs expressed by the customer concerning the desired product or service, its availability date, and required support. The performer must also identify requirements that may not have been expressed by the customer, but which are needed to accomplish the contract and it must identify legal or regulatory requirements. Identification, review, and communication of customer requirements can be achieved in a single process—comprehensive contract review.

2.2.2 Contract Review

The formal relationship between a customer and a performer begins with a contract. It cannot be stressed too often that the contract defines how well a job will be done. This is so because a properly written contract in which the requirements and inherent characteristics of the product or service are expressed is the very definition of what the customer is going to get. When supported by ongoing reviews with the participation of the customer, the execution of the contract will meet the customer's expectations. Some expectations will be unexpressed and hence not in the contract, but if it becomes apparent that they are not being met, the contract as written becomes inadequate and should be amended. Amendment is not so much a correction as it is a continuation of the contracting process.

Peter Hybert (1996) defines contracting in a way that enhances customer satisfaction, which is the dominant trend in today's intense global competition: "Contracting is the process by which systems are designed and delivered." This definition goes much further than convention dictates. Most people will agree that the contract is not over until delivery, but Hybert is saying more than that. He is saying that *contracting* is not over until delivery. This notion carries with it the sense of an ongoing process. It goes beyond the conventional idea that contract review is a single event in the beginning of a contract phase.

An effective contracting process will track customer requirements and expectations during the entire period of performance. Both may change as a result of developments that can occur during an extended period of performance. Figure 2.1 depicts the steps of an effective contracting process. The first step is to identify customer requirements, which usually starts with an initial meeting of customer and performer. At that time, an idea of customer expectations is established.

In simple cases a single person may represent each player in the contract. For example, a customer-agent can express exactly the requirements for an

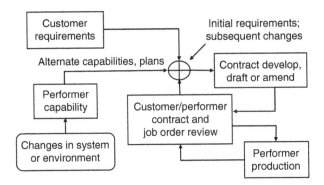

Figure 2.1 An effective contracting process. *Source:* From *The Role of Sarbanes-Oxley and ISO 9001 in Corporate Management: A Plan for Integration of Governance and Operations* © 2012 William A. Stimson by permission of McFarland & Company, Inc., Jefferson, NC. www.mcfarlandpub.com.

off-the-shelf product and a single performer-agent can take the order. The performing company must determine which of its activities is to be the primary interface agent with the customer. Companies differ on this. Sometimes the marketing department is primary contact. In others, the customer service department is the point of contact. Still other companies may have a sales office that serves as primary contact with the customer.

However, if the product were a large system or complex, or if the service were a major project, then the contracting process would consist of teams, one representing the customer, the other the performer. For example, if Boeing Aircraft Company were the performer and American Air Lines the customer, each side would present a team of experts representing the various concerned activities within each corporation. The performer experts would meet with the customer experts to identify requirements and to determine whether those needs could be satisfied by company capability.

The performer's team will have members from marketing, purchasing, design, sales and service, and manufacturing. The customer's team will consist of experts of the various capabilities that the system will provide. For example, there may be personnel from materials requirements planning, information technology, maintenance crews, operators, planning, and human resources. A customer purchasing a passenger airplane will want people in the contract review process who represent its own marketing, service, technical support, and maintenance functions and who understand their customers' requirements.

Once customer requirements are identified, the next step is to review the capability of the company to meet the requirements with its existing facilities or to determine if new processes are necessary and feasible. Assuming there is an initial agreement of expectations between customer and performer, a contract is developed and an initial review takes place, including the job order if the customer wants to see it. In some cases, the contract contains the customer requirements and not the specifications. In other cases, the contract will contain both. In all cases the people in operations work from the job order. The customer may verify from the contract that the specifications meet the requirements, but this does not verify the job order, which may contain more, fewer, or different specifications.

The contracting process is structured to maintain an agreement of expectations between the performer and the customer throughout the performance period. This is shown in Figure 2.1 as a feedback element of the system. It lets the customer verify that requirements were recorded as they were transmitted and that the specifications are a valid translation of the requirements. During the performance period, it provides the customer and the performer the occasion to review progress and resolve developing problems together. This helps the customer to understand what must be done, the options available, and their cost, all of which may vary during the performance period because of external forces. The customer's expectations may rise or fall as a result of the review, which enhances an agreement and satisfaction at the end of the contract.

Figure 2.1 shows that after an agreement on the contract is reached, the company begins the various phases of work: design, development, and production, the periodic contract reviews continue throughout the process. During these reviews, both customer and performer may request changes. Perhaps a material is no longer available, or its price has increased, which may change the scope or the cost of the product. Perhaps the customer's needs have changed. Joint review by customer and performer enhances the opportunities for maintaining an agreement between them of customer expectations. The following are some of the events that are considered in the contract review process:

- An initial meeting with the customer is conducted to establish the product or service requirements, including those for delivery and support, if any;
- Customer requirements are balanced against company capabilities and resources, and against regulatory requirements;
- After a contract is drafted, an initial joint review is conducted to ensure that the requirements are adequately defined and confirmed, and specifications documented;
- An ongoing process of periodic contract and performance review takes place in which the customer is a participant, either directly or through continual feedback;
- Customer satisfaction with the final product or service is assessed and recorded.

2.3 Specifications

Specifications are documents of prerequisite information that are necessary to accomplish correctly the tasks, processes, products, and services required by customer requirements. Often simply called "specs," they specify functional or design data, usually quantitative, in the performance or creation of a material, design, product, service, or process.

Prerequisite information defines the required characteristics of a product, service, or procedure. That is, how a thing is to be done. Such documents include but are not limited to proposals, technical standards and instructions, technical and performance manuals, policy documents, performance agreements between performer and customer and between contractor and subcontractors, operational procedures, plans, drawings, emails, and change orders.

However, in forensics, post-hoc data share equal importance to prerequisite data and hence comprise a second category of specifications. Post-hoc information describes the characteristics of a finished product, service, or procedure. That is, how a thing was actually done. Such documents include but are not limited to test and inspection reports, performance and vendor evaluations, reviews, memoranda, nonconformity reports, status reports, engineering change proposals, audit reports, and emails.

In the design phase of engineering, there are always quantitative criteria that describe critical-to-quality characteristics of the product or service to be manufactured or provided. For example, the design of an engine will include torque criteria. The design of a cafeteria will include queuing time criteria. Design criteria consist of a target value and limits above and below the target value that are relevant to the quality characteristic of the product or service. The limits are established in recognition that infinite precision is not possible. A good deal of thought is used in determining these limits, along with recognition that they may have to be changed. For example, the design may specify that a cylinder should be 3 inches in diameter, ±0.0001 inch, but the engineers may then find that the production system cannot meet these "specs." Sometimes a degree of intuition is used in writing specifications. If the production system is not capable of meeting the criteria, there are only two choices: widen the specs or buy a new production system.

Widening the allowable limits of the specification increases customer risk. Conversely, it is expensive to buy new production equipment. A dilemma of this type challenges objectivity and introduces the notion of ethics. This kind of situation provides another good reason why the customer should be in the contract review process.

The ability of a performing system to meet customer requirements is called capability. This is formally defined as how much of the product variability is bounded within the specifications. Usually, this means that the inherent variation of the productive process is within specification limits. A process can be stable and not be capable. The variability of the process may well be bounded, but if that variability exceeds the specification, then the process is not capable. You need to design a new process. A process must be stable before its capability can be determined. If the product or service is quite expensive, the wise customer will ask to see evidence of process stability. Ex post facto, the forensic systems analyst will be vitally interested in the stability of any process during the period in litigation.

2.4 Standards

Standards provide uniformity in performance. For example, measurement standards allow the same measurement to be made of the same parameter irrespective of conditions, time, or place. Measurement is always defined in terms of comparison to a standard. Whether the measurement is one of space, weight, management, behavior, virtue, or performance, there must always be a standard of comparison. Notice that the term measurement is used here in its most general sense, from measuring the length of a field to measuring up to one's responsibilities. All measurements must have a standard.

This book is concerned with standards of management performance. Table 2.1 provides a list of a few performance standards that give the reader a sense of the scope of such standards. They cover a wide spectrum of management activity: information technology (IT), ethics, business risk, governance, and quality. IT is a special concern because it invariably includes a company's financial system. All of these activities come under the purview of industrial standards. Nonetheless, all the sponsors are private organizations. In the American free enterprise system, business conduct is usually regulated by professional or private groups and not by government.

Table 2.1 A partial list of performance standards and their sponsors.

Standard	Sponsor
Principles of Corporate Governance	Organization of Economic Co-operation and Development
OCEG Framework	Open Compliance and Ethics Group
Policy Governance Model	International Policy Governance Association
Enterprise Risk Management Framework	Committee of Sponsoring Organizations of the Treadway Commission
IT Control Objectives for Sarbanes–Oxley	IT Governance Institute
ISO 9000 Quality Management Standard	International Organization for Standardization

Source: From The Role of Sarbanes-Oxley and ISO 9001 in Corporate Management: A Plan for Integration of Governance and Operations © 2012 William A. Stimson by permission of McFarland & Company, Inc., Jefferson, NC. www.mcfarlandpub.com.

I once attended a national meeting of Navy personnel whose task at hand was to write standard work items (Standard Specifications for Ship Repair and Alteration Committee, 1984). In one effort, negotiations broke down completely over an argument of what "to do" meant. This type of argument is common in anything concerning contracts and standards, and the whole purpose and destiny of a standard is to appear in a contract.

Wisdom dictates that every standard should begin with definitions of key words that have specific meaning to the performance required in that standard. Even very common words can be given specific meaning for performance purposes. As an example, the word "controllable" has a specific meaning to engineers, but it may mean something else entirely to the general public.

In another example, the word "durability" is sometimes used by both manufacturer and customer to describe how long in time a product or service may be relied upon. In common language, *durability* and *reliability* are often interchangeable. To engineers, they are not at all equivalent. *Durability* has no engineering or technical meaning. It can mean whatever you want it to mean, which may be quite different to the understanding of another party. Conversely, *reliability* has a very rigorous meaning to engineers: it is the probability that an item will perform a required function without failure under stated conditions for a specified period of time. Hence, once the function, conditions, and specified time are defined, product reliability becomes a specification. To the forensic analyst, product reliability may well be a critical issue in litigation. "Durability" is to be avoided as a defense or complaint because technically it means nothing.

A final example clarifies the importance of mutual agreement on the meaning of words in a contract. A standard might define the word *supervise* as "to lead, oversee, and inspect the work of others." Then at some later time in a contract in which the standard is invoked, the performer may be required to supervise a given task, and it will be clear to customer and performer what must be done from a legal point of view.

A properly written standard will tell the performer what to do, but never how to do it (Office of Management and Budget (OMB), 1998). There is more at stake here than offending performers by implying that they need to be told how to do a task. If you tell the performer how to do the job, you own the result. If things go badly, the performer's plea will be that "we were just following orders." The wording must be unambiguous not only to keep performers from being confused but to keep them from claiming to be confused.

The story a few paragraphs earlier about the three-person management teams is a good example of how a poorly worded standard permits deliberate misunderstanding. We wanted at least three managers. We meant that if the job required five or six, then we expected that the greater number would be assigned. However, the contractor understood that logically speaking, *exactly three* satisfies the requirement of *at least* three, and therefore, although inadequate, the lesser number was a legal solution to the problem and a lot cheaper. The goal was to maximize profit even if it meant risking good performance.

It is all well and good to talk about customer satisfaction and the meeting of customer expectations. Most performers profess to do so, or at least to try. However, business works on the expectation of profit, so rarely do customer and performer have the same expectation. Well written contracts, specifications, and standards are very important in the prevention of misunderstanding and disappointment of either customer or performer. Often, delicate situations arise in which ethical issues get involved in contracts.

Forensic systems analysts are rarely experienced in contracts, nor are they hired for that purpose. However, the attorneys that employ them will be well versed in the facts, phrases, and nuances of contracts. The two—analyst and attorney—will work together to determine whether a finding by the analyst is relevant to the requirements of the contract. You get what is in the contract, nothing more.

Credits

This chapter, edited to the purpose of forensic systems engineering, is taken from *The Role of Sarbanes-Oxley and ISO 9001 in Corporate Management: A Plan for Integration of Governance and Operations* © 2012 William A. Stimson by permission of McFarland & Company, Inc., Jefferson, NC. www.mcfarlandpub.com.

References

Hybert, P. R. (1996). "Five Ways to Improve the Contracting Process." *Quality Progress*, February, pp. 65–70.

Office of Management and Budget (OMB) (1998). *A Guide to Best Practices for Performance-Based Service Contracting.* https://obamawhitehouse.archives. gov/omb/procurement_guide_pbsc/. Accessed September 13, 2017.

Standard Specifications for Ship Repair and Alteration Committee (1984). *United States Navy Standard Specifications for Ship Repair and Alteration Committee.* San Francisco, CA: United States Navy. The standard was Standard Item 009-67 (FY 86), dated September 1, 1985 and appeared in all contracts requiring combat systems alterations.

United States Naval Sea Systems Command (1989). *NAVSEA Standard Work Item 009-67.* http://www.navsea.navy.mil/. Accessed September 19, 2017.

3

Management Systems

W. Edwards Deming (1991) and Joseph Juran (1992) estimate that 80–94% of system problems are the responsibility of management. The forensic approach to system analysis is then quite clear—system effectiveness is greatly influenced by its management system. In French, management is called *la direction*, a term I like because it describes exactly what it is that management is supposed to do. One of the most important tasks of management is to ensure the effectiveness and efficiency of an organization's processes. This is a comprehensive task and applies to all kinds of processes: education, government, production, service, commercial, military, and so on. The task includes the management framework to be used to establish and maintain policies, goals, resources, procedures, processes, and effective performance. In short, the task is to define and establish a management system.

Forensic Systems Engineering: Evaluating Operations by Discovery,
First Edition. William A. Stimson.
© 2018 John Wiley & Sons, Inc. Published 2018 by John Wiley & Sons, Inc.

3.1 Management Standards

3.1.1 Operations and Good Business Practices

Operations can be managed by "the seat of your pants" or they can be managed in some orderly fashion. The first kind of management is reactive and leads to endless fire drills and panic mode. The second is proactive and uses a systematic approach through recognized good business practices. In this book, the systematic approach is called a management system. A company with an effective management system has the ability to consistently meet or exceed customer expectations in the goodness of its products and services.

A management system should be synergistic—the parts work together effectively to achieve the system goal. In practical terms, all the productive and supportive activities in the company are integrated and coordinated to achieve corporate goals. All productive processes should be organized in the natural flow of things and supported with necessary resources. This type of structure is called the process approach and is suitable to the newer international management standards such as ISO 9001. Also, the performance of the management system itself is continually measured for effectiveness and efficiency.

There is nothing mysterious or abstract about good business practices. They have been identified for every industry and serve as benchmark processes. Some are specific, such as those used by process industries; some are general, such as those developed for guidance in management. Models derived from good business practices can become *standards* that are used to organize management systems. The standards and systems complement one another as form and substance. The model provides the form; the system provides the substance. This point is important to system design because it means that you cannot have systematic performance without a standard operating procedure.

Substance is created from form, which is embedded within it. Thus, where you have both form and substance the two are inseparable, but form precedes substance. For example, the Constitution is our model of government, from which the functioning structure was built. Therefore, we begin the discussion on management systems with an analysis of management standards—the model of the system.

3.1.2 Attributes of Management Standards

An effective standard of operations management must have four attributes. The effective standard (i) provides uniformity in performance; (ii) attracts a large number of subscribers sufficient to represent the industry; (iii) is physically realizable; and (iv) has legal status to protect against liability. The first attribute, uniformity in performance, is achieved by judicious wording of each requirement so to serve as a control of the objective activity. Properly written, clear, and unambiguous, a standard ensures uniformity in performance because deviation is unlikely, given common understanding of what is to be done.

A standard is, by definition, an agreement among participants to do something in a particular way. If the requirements of the standard were so difficult as to make profitable enterprise unlikely, there would be few players. Hence, a standard is a compromise because it must appeal to a large group, all volunteers. It will be acceptable to a given population if they see some benefit to it. Thus, the second attribute of any standard is that its terms are agreeable to subscribers.

The third attribute of a standard is that it is realizable—blue sky standards cannot be implemented. Every management system has two parts—the documented part and the physical or implemented part. A company can easily create a documented system that is compliant to a given standard, but the question is whether it can be implemented and whether it will be effective. Even if the standard is realizable, the physical system may be incorrectly implemented, which is why forensic systems analysts must verify both parts of the system. But if the standard is not realizable and the documented system duplicates it, the implemented system cannot work.

The fourth attribute of a standard is that it has legal legs—it can stand up in court. Randall Goodden (2001) says that for a management system to protect a company against product liability, it must have a fully documented system of control procedures. Control procedures support the claimant of good business practices and standards of care. A comprehensive set of documented control procedures usually resides in the standard that defines the system.

3.2 Effective Management Systems

In principle, every business activity in a free society is permitted by law to operate as it chooses, subject to those constraints deemed by the government as necessary to protect the public. As I argued previously, business operations can be effective if they follow systematic procedures deemed by industry itself to be good business practices. Hence, the forensic systems analyst can expect to find some governing management system in place in any investigation. Each company is free to choose its own system except if performing on a contract that specifies a particular system. Usually, the management system will be a variant of one with public recognition. Some of the more prominent systems that might be employed are a Malcolm Baldrige program, Total Quality Management (TQM), Six Sigma, Lean, and the Production Part Approval Process (PPAP).

3.2.1 Malcolm Baldrige

The US Congress established the Malcolm Baldrige National Quality Award (MBNQA) program in 1987, with Public Law 100–107. The purpose of the program is to establish the global leadership of American industry through continuous improvement of methods of operations. The Baldrige Performance Excellence Program (2015) criteria are comprehensive and provide criteria in leadership; strategic planning; customer focus; measurement, analysis, and knowledge management; human resource focus; process management; and performance results. An applicant company is free to meet the criteria in any way it chooses, and the annual award goes to the best competitor in six business sectors: manufacturing, service, small business, education, health care, and nonprofit/government.

An MBNQA program is clearly a model operations management system and a very good one at that. However, it is an award, not a standard. The criteria are strict and only the most determined and dedicated companies will put the required effort into going after the award. If the MBNQA were adopted as an operational standard, it is not likely that the average company would pursue it any more intensely than they would pursue the award. Recall what I said about standards— if they are too tough, you have few players.

However, even discounting its capacity as a standard, an MBNQA program has much to offer. It can be used as a freelance model—you use as much of it as you want, where you want. In this sense, the program can become a formal operations management system of arbitrary design, using a model adapted from the Baldrige criteria.

3.2.2 Total Quality Management

TQM, as it is popularly known, is used widely but the way it is implemented varies from user to user. It is a freelance system with some common principles. Oppenheim and Przasnyski (1999) claim that TQM can be an effective and comprehensive management system for operations. However, TQM is not a model and has no standard form. Whether a company would have a system of documentation as a defense in liability litigation would be entirely up to its own ingenuity—nothing in TQM protects either the company or the customer in legal redress. Without some fixed format it would be difficult to demonstrate standard good business practices or business standards of care. TQM has no legal legs and so fails an attribute required for an operations management standard.

3.2.3 Six Sigma

Extremely popular in business, Six Sigma is a strategic management methodology (GE Fanuc Company, 1998). Roger Hoerl (2001) writes that it is "One of the few technically oriented initiatives to generate significant interest from business leaders, the financial community, and the popular media." This interest is owed, no doubt, to its focus on the corporate bottom line. Every project adopted for Six Sigma activity requires a projection of financial return.

The name "Six Sigma" is a statistical term that measures how far a given process output deviates from its design value. The objective of a Six Sigma program is to implement a measurement-based strategy that focuses on process improvement and variation reduction using statistical techniques applied to improvement projects. Thus, Six Sigma is project oriented. The program is formal, employing defined algorithms for process design and for problem solving.

The forensics systems analyst should be aware that there are a few caveats to a Six Sigma program that are not necessarily shortcomings but can easily degenerate to such. The first is the fundamental nature of project orientation. Project thinking brings both strengths and weaknesses. The strength is obvious— every process in the company can focus on one problem after another, always moving toward improvement and always recording financial gains, because a Six Sigma program requires financial improvement goals. The downside is that Six Sigma is a great de-optimizer. As Eliyahu Goldratt (1990) explains, a system has both subsystem and total system constraints and they are usually not the same. Therefore, if each subsystem attempts to optimize itself, the total system will be less than optimal.

The project approach is the converse of the strategic approach—it is tactical. Both are necessary to long-term success, but there is little that is inherent in Six Sigma to address the big picture. To be strategically useful, a Six Sigma program must integrate with an existing strategic and visionary structure. Fundamentally, it is a problem-solving methodology.

The last caveat is particularly relevant to the utility of Six Sigma as a standard of operations management. Companies don't get certified in Six Sigma—individuals do, but there is no central certifying body. Many companies, resisting the high cost of training, may pay for the certification of one employee, and then train the remaining team in house. This is freelance certification. Therefore, as effective and efficient as Six Sigma might be as a problem solver, it cannot serve as a management system standard. The Six Sigma measuring stick for effectiveness is project goal achievement and financial benefit.

A Six Sigma program can be used as a freelance model—you use as much of it as you want, where you want, when you want. You get the power of Six Sigma to solve problems and gain improvements, and you can superimpose this capability on any formal operations management system.

3.2.4 Lean

Once you get past the revulsion of using "Lean" as a noun, you find it an efficient manufacturing process. The technical community has long shown a disdain for the English language and this is not my first battle of this kind. I lost the first one too. Joining IBM as a systems engineer, I had to learn about "initializing" the system, a term I hated. It turns out that initializing the system is extremely important in system testing and is not the same as "beginning" or "restarting" the test again. To "initialize" a test means to restart it again only after resetting initial conditions. I just wish they could have found a better word.

Lean is an American derivative of the Toyota Production System (Liker, 2004). It focuses on waste and cost reduction, achieving continuous flow and pull-production through incremental and breakthrough improvements. To get a flavor of Lean, consider that an important measure of Lean effectiveness is *touch time*, the amount of time a product is actually being worked on. Lean is definitely a tactical management system.

A study made by David Nave (2002) compares the improvement programs of Lean and of Six Sigma, and his comments are particularly useful in evaluating these methodologies as operations management systems. In an example, the author states that one drawback of Six Sigma is that system interaction is not considered and that processes are improved independently. This comment reinforces my earlier claim that Six Sigma is often a de-optimizer. Nave then gives as a Lean shortcoming that the method is not very strong on statistical or system analysis. Altogether, the author's comments raise questions about the effectiveness of either method as a strategic management system.

My own observation is that Lean is often too lean. Its basis, the Toyota Production System, assigns a set of tasks in three categories: value added, non-value added, and non-value added but necessary. Validation and verification fall into the third category. In the United States, however, all too frequently Lean consists of only two categories: value added and non-value added, with validation and verification tasks assigned to the second category. Thus, if the production schedule slips, test and inspection are truncated or abandoned altogether. Yet, if the customer complains, the performer asserts that the company is using Lean, or more sanctimoniously, *LeanSixSigma* (as though the title were sufficient to command respect). Unfortunately, they are being lean, but too lean. This misconception, or misfeasance, remains to be clarified in the courts.

3.2.5   Production Part Approval Process

PPAP is a standardized process in the automotive and aerospace industries that helps manufacturers and their suppliers to communicate and approve production designs and processes during all phases of manufacture. PPAP helps ensure that the processes used to make parts can consistently reproduce them at stated production rates during regular production runs. In the automotive industry, the PPAP process is governed by the Automotive Industry Action Group, the original developer of the process.

The automotive industry can be characterized by its vast supplier network. One wag has commented that the United States does not make cars anymore, she assembles them. The PPAP was developed to solve the problem of supplier control. Important elements of the process include design records, engineering change orders, customer approvals, defined and documented processes, and process interconnections. All subject processes are monitored, analyzed, and subject to continual improvement. Records of each activity and step are created, maintained, and retained.

The aerospace industry, too, depends heavily on suppliers; however, there is no standardized PPAP in the aerospace industry, with each company developing its own requirements. In addition, the industry relies frequently on the First Article Inspection process for PPAP requirements on dimension measurements. This latter process is discussed in Chapter 16.

The PPAP is clearly a standard and a good one at that. Its focus, however, is on supplier control and documentation. It could be broadened and adapted to general management, but there is a competitive trend against it—the adaption of ISO standards to the automotive industry. For example, the International Automotive Task Force (IATF) has published IATF 16949 (2016) as the quality management standard for the automotive industry; AS9100C (2015) serves as the quality management standard for the aerospace industry.

3.3 Performance and *Performance*

In the competitive global economy, there is a need for a performance standard that can deliver effectiveness and efficiency in its processes, quality of its output, and profits. The latter cannot be overestimated. In their book, *The Goal*, Eliahu Goldratt and Jeff Cox (1986) assert that the goal of a (manufacturing) company is to make money. However, in this book, we recognize that there are really two independent goals: profit and customer satisfaction. Goldratt and Cox have their eye on only one sparrow.

I can make a fair case that ours is a litigious society. If so, we must have a standard of performance that contains internal controls sufficient to establish culpability or innocence in process liability. I believe that the international management standard, ISO 9000, is such a standard. It is frequently required in government contracts of performance and is increasingly recognized in our courts as a measure of prudent management.

3.4 Addendum

Several times in this chapter, I used the term "freelance" without defining it. Perusing through the chapter, readers might have thought of freelance as self-assessment and benchmarking. To distinguish these terms, I define them this way. *Freelance* means to create your own system. *Self-assessment* means to refer to, or "benchmark" to, a standard or a well-defined system and then evaluate the compliance of your own system to the model. In principle, there is nothing wrong with either strategy. The forensic systems analyst is concerned that the implemented model is compliant to the reference standard and that it is effective.

Credits

This chapter consists of various sections taken from *The Role of Sarbanes-Oxley and ISO 9001 in Corporate Management: A Plan for Integration of Governance and Operations* © 2012 William A. Stimson by permission of McFarland & Company, Inc., Jefferson, NC. www.mcfarlandpub.com.

References

AS9100C (2015). *Quality Management Systems—Requirements for Aviation, Space and Defense Organizations.* Washington, DC: Society of Automotive Engineers (SAE) International. http://standards.sae.org/as9100c/. Accessed September 19, 2017.

Baldrige Performance Excellence Program (2015). *2015–2016 Baldrige Excellence Framework: A Systems Approach to Improving Your Organization's Performance.* Gaithersburg, MD: National Institute of Standards and Technology. https://www.nist.gov/news-events/news/2014/12/2015-2016-baldrige-excellence-framework-and-criteria-businessnonprofit. Accessed September 20, 2017.

Deming, W. E. (1991). *Out of the Crisis.* Cambridge, MA: Massachusetts Institute of Technology, p. 315.

GE Fanuc Company (1998). *Six Sigma Quality,* GE Fanuc brochure. Charlottesville, VA: GE Fanuc Company.

Goldratt, E. (1990). *What Is This Thing Called Theory of Constraints and How Should It Be Implemented?* New York: North River Press.

Goldratt, E. and Cox, J. (1986). *The Goal: A Process of Continuous Improvement.* New York: North River Press.

Goodden, R. (2001). "How a Good Quality Management System Can Limit Lawsuits." *Quality Progress,* June, pp. 55–59.

Hoerl, R. (2001). "Six Sigma Black Belts: What Do They Need to Know?" *Journal of Quality Technology,* October, pp. 391–406.

IATF 16949 (2016). *Automotive Quality Management System Standard.* International Automotive Task Force. http://www.iatfglobaloversight.org/iatf-publications/. Accessed September 19, 2017.

Juran, J. M. (1992). *Juran on Quality by Design.* New York: Free Press, p. 428.

Liker, J. K. (2004). *The Toyota Way.* New York: McGraw-Hill, p. 7.

Nave, D. (2002). "How to Compare Six Sigma, Lean, and the Theory of Constraints." *Quality Progress,* March, pp. 73–78.

Oppenheim, B. and Przasnyski, Z. (1999). "Total Quality Requires Serious Training." *Quality Progress,* October, pp. 63–73.

4

Performance Management: ISO 9001

The major assumption of this book is that there is a causal relation between a process and the product or service resulting from this process. Thus, the operation of the process and its performance are essential to forensic systems analysis. The management of performance is a critical component of operations and may well be the focus of forensic investigation. Forensic analysts, fundamentally auditors, will compare what is to what should be, which implies that a standard of performance is required—indeed, it is a tool of the trade. The most commonly used such standard is the international standard, ANSI/ISO/ASQ. Q9001-2015. *American National Standard: Quality Management Systems— Requirements*, also known simply as ISO 9001.

ISO 9001 is not without controversy. Across the nation, you find blue banners stretched across the porticos of business after business, proclaiming their certification. Across the nation, you find article after article in the technical journals denouncing the Standard as an expensive and impotent charade. Critics claim that ISO 9001 is not specific enough to provide performance excellence. Champions of ISO 9001 reply that, on the contrary, the failure to be specific is actually its strength. ISO 9001 was written to be applied to the widest spectrum of business and is necessarily general in form. Because of this generality, it is easily adaptable to both strategic and tactical purposes.

Forensic Systems Engineering: Evaluating Operations by Discovery,
First Edition. William A. Stimson.
© 2018 John Wiley & Sons, Inc. Published 2018 by John Wiley & Sons, Inc.

This book will not settle the matter. What is essential to forensic investigation of operations is that some recognized and acceptable standard of performance is used and that a very large number of companies and industries use ISO 9001 as their quality management system (QMS) standard. Many government contracts require its employment. Sooner or later, if you are a forensic systems analyst, you will be required to render an opinion on the effectiveness of a given system in a court of law and you will need a standard of reference. Arbitrarily, ISO 9001 is the standard chosen for this book.

4.1 Background of ISO 9000

The ancestry of ISO 9000 begins with World War II. Coalitions on both sides of the war needed means of achieving uniformity of mass-produced product. Production levels attained historically unimaginable heights and product quality could no longer be assured using traditional product standards. Wartime research brought advances in systems analysis and operations research that could be applied to production methods. This research led to the use of methods to analyze the process rather than the product to ensure resultant quality. The basis of process analysis is the belief that if the process is good, the product will be good too, even in mass production. Some of the early research, such as that of Walter Shewhart (1931), had been done in years prior to the war, but gained major impetus under wartime demands.

Therefore, World War II brought about the development of quality management systems that were composed of an inspection scheme augmented by a program of quality assurance (QA) that acted independently over operations. A classic example of this kind of system, used for over 50 years throughout the defense industry, was defined by the standards Mil-Q-45208 (1963) and Mil-Q-9858A (1993), which pertained to inspection systems and QA systems, respectively. Readers who are familiar with ISO 9001 would find similarities in Mil-Q-9858, which required procedures for employee empowerment, contract review, documentation control, and production control.

After the war, the world economy began a slow transition to a single marketplace. In 1947 the General Agreement on Tariffs and Trade (GATT) was signed, which defined standards of trade and production. In accord with this trend, the European Economic Community was created in 1979, one in a series of free trade groups culminating in today's European Union. In 1946, the International Organization for Standardization (abbreviated "ISO") was established in Switzerland and assumed oversight of the standards of various member nations. ISO publishes many different kinds of standards, among them standards for environmental management and medical devices. One of its most utilized standards is ISO 9000, *Quality*

Management Systems, first issued in 1987, which is concerned with the effective performance of process management.

4.1.1 ISO 9001 in the United States

In the 1970s, Japanese electronic and automobile manufacturers began to achieve deep inroads in the American market and US manufacturers were obliged to recognize the reality of a global economy. In response, many companies adopted ISO 9001 as their standard for managing operations. Those companies with an interest in exporting their products had little choice, as ISO 9001 certification became a market requirement in many industrial nations.

In the United States, administration of the ISO 9000 standards program is assumed by the American National Standards Institute (ANSI) and by the American Society for Quality (ASQ), who jointly accredit the program through the ANSI–ASQ National Accreditation Board (ANAB). The ANAB accredits private companies as certifying bodies, who are then authorized to certify manufacturers and providers of service to any of several standards.

Certification to ISO 9001 at the time of this writing is about 33,000 companies in the United States, including some Federal agencies such as the Naval Surface Warfare Center at Carderock, MD. Internationally, over 1 million companies are certified (ISO, 2016).

4.1.2 Structure of ISO 9000:2005

The name "ISO 9000" is somewhat ambiguous. It is properly used in two ways. First, ISO 9000 is a set of standards for quality management systems, and secondly, it is the first standard in the set, which includes, as of this writing, the following:

- ISO 9000:2005. *Quality Management Systems—Fundamentals and Vocabulary* (ANSI/ISO/ASQ, 2005)
- ISO 9001:2015. *Quality Management Systems—Requirements* (ANSI/ISO/ASQ, 2015)
- ISO 9004:2009. *Quality Management Systems—Managing for Sustained Success of an Organization—A Quality Management Approach* (ANSI/ISO/ASQ, 2009)

The reader will note that each standard has a different date. This condition comes about because each standard is revised according to demand, and in any case, the forensic systems analyst will be concerned only with a standard whose date of issue corresponds to the period in litigation. As many trials go on for years, a standard appropriate to a given case may not be the most current.

Although certification to ISO 9001 is formal, it is not mandatory. A company is free to forego certification, ignore the Standard, or simply use it as a model, implementing it as the company sees fit. This is called self-assessment and carries no recognition beyond the company. However, many companies choose to obtain certification, granted by the certifying bodies operating under the auspices of the ANAB.

Standards ISO 9000 and ISO 9004 are advisory. Only ISO 9001 contains contractual requirements and, perhaps for this reason, many companies implement only this single standard and more or less ignore ISO 9004, which plays an advisory role on achieving excellence. This narrow view is regrettable and a missed opportunity. It may also be at the heart of a long held criticism: ISO 9001 is strong on form and weak on substance, because the effectiveness of the Standard is not inherent but depends upon how well it is implemented. Performers who certify to ISO 9001 and ignore ISO 9004 are taking the path of least effort.

Moreover, the contractual status of ISO 9004 may be underestimated. In the United States, guidance standards such as ISO 9000 and ISO 9004 are viewed as components of a series along with ISO 9001 that can be used to examine issues such as product safety. Guidance documents that are part of a series can be used to establish a company's "due diligence" and "duty of care" and can be used by the courts to establish evidence of negligence (Kolka, 1998). Though ISO 9004 is not contractual, a plaintiff could argue that it should be reflected in the QMS of any company registering to ISO 9001 as a set of good business practices. We shall see in Chapter 8 that such an argument has carried weight even when the defendant had no obligation to ISO 9001.

According to Jack West Chairman of the US Technical Advisory Group 176 and lead American delegate to the International Organization for Standardization, an enterprise needs both ISO 9001 and ISO 9004. The first standard, ISO 9001, provides the form for governance. The second standard, ISO 9004, provides a performance excellence model that can make a company a world-class competitor. This is the substance of the ISO 9000 set of standards. West calls the two standards a "consistent pair" that will enhance market success (West et al., 2000). Thus, ISO 9001 can be both a standard of governance and, if fully complemented with ISO 9004, a standard of performance excellence.

4.1.3 The Process Approach

All of us are artists in that we usually carry around in our mind images of how things work. Engineers are no exception and I often picture the flow of a process in my mind. Figure 4.1 shows a model of a management system for operations. The model is very general—any effective management system is structured similarly—and is quite useful for forensic analysis of systems. ISO 9001 is also structured in this way, with seven core requirements configured in a process approach—the natural flow of operations. The requirements in

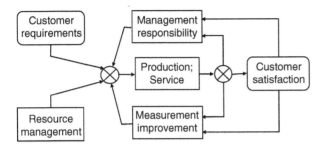

Figure 4.1 A model of a management system for operations. *Source*: From The Role of Sarbanes-Oxley and ISO 9001 in Corporate Management: A Plan for Integration of Governance and Operations © 2012 William A. Stimson by permission of McFarland & Company, Inc., Jefferson, NC. www.mcfarlandpub.com.

this flow chart are listed by numbered clauses. In the 2015 version of ISO 9001, the clauses are Clause 4, *Context of the Organization*; Clause 5, *Leadership*; Clause 6, *Planning*; Clause 7, *Support*; Clause 8, *Operation*; Clause 9, *Performance Evaluation*; and Clause 10, *Improvement*.

The requirements are expressed in the language of top management and pertain to those processes that affect the effectiveness of operations. The forensic systems analyst should not take a narrow view of operations as only that function defined by Clause 8. Relative to the effectiveness of the QMS, *all* the clauses affect operations because all subsystems interact synergistically. The clauses and the sub-clauses that go with them describe the scope of requirements. Actually, they do more than that. They are instructive in that they tell you what kinds of policies and procedures should be going on in each clause. They are written in terms that are sufficiently general to accommodate the strategic policies and the tactical procedures of various activities across the corporation. For example, Clause 5.3, *Organization Roles and Responsibilities*, describes obligations to customers, employees, and shareholders. It is here that shareholder policies and procedures could be spelled out.

Table 4.1 is an outline of the core requirements whose accomplishment can be measured. A few of the subtitles have minor changes in order to fit them into the table. A cursory look at the figure shows that ISO 9001 is indeed descriptive of the totality of operations. Study of the requirements shows that the Standard is a comprehensive management model, spanning operations, resource management, and design.

Unfortunately, the clause numbering system of the latest version of ISO 9004 (ANSI/ISO/ASQ, 2009) is no longer correlated to that of ISO 9001 and introduces an unnecessary confusion among users and on the forensics systems analyst to associate the two. As I said earlier in this chapter, guidance documents that are part of a series can be used to establish a company's due diligence and can be used by courts to establish evidence of negligence. The forensic

Table 4.1 Quality management system requirements of ISO 9001.

4.0: Context	4.1: Organization; 4.2: Needs and expectations; 4.3: Scope; 4.4: QMS and processes
5.0: Leadership	5.1: Commitment; 5.2: Policy; 5.3: Roles, responsibilities, and authorities
6.0: Planning	6.1: Risks and opportunities; 6.2: Quality objectives; 6.3: Changes
7.0: Support	7.1: Resources; 7.2: Competence; 7.3: Infrastructure; 7.4: Environment; 7.5: Monitoring and measuring; 7.6: Organizational knowledge
8.0: Operation	8.1: Planning and control; 8.2: Requirements; 8.3: Design and development; 8.4: Supplier control; 8.5: Production/service provision; 8.6: Release; 8.7: Control of nonconforming products
9.0: Performance evaluation	9.1: Monitoring, measurement, analysis, evaluation; 9.2: Internal audit; 9.3: Management review
10.0: Improvement	10.1: General; 10.2: Nonconformity and correction; 10.3: Continual improvement

systems analyst must somehow correlate ISO 9001 and ISO 9004 in order to arrive at a clear understanding of the effectiveness of operations in litigation.

Fundamentally, the Standard requires that the factors governing quality of product are under control and that the process is documented. The details of implementation are left to individual companies on the grounds that each company has its own way of doing business. The scope of the Standard is defined by its requirements that are applicable to a particular company, and so depends upon the breadth of operations of that company. For example, a company that provides no service or does no design will have less breadth of operation than one that does and will have fewer ISO requirements upon it. This flexibility is known as exclusion. Processes that are defined in the Standard but which do not exist in the company, or that do exist but will have no effect on customer satisfaction with the product or service, may be excused from compliance.

Among the criticisms leveled against ISO 9001 is the seemingly great amount of required documentation. But from a legal point of view, documentation is a major asset of ISO 9001 because it provides records and describes internal controls. There is a cycle of paperwork in all business transactions: sales orders, purchase orders, job orders, and delivery orders, and so on. Figure 4.2 shows this cycle as it might apply to manufacturing. Various records are used in this cycle for customer requirements, design specifications, parts and materials, fabrication and assembly, test and inspection, and handling and packaging. This documentation defines a paper trail from customer expectations to delivery. If customer dissatisfaction leads to litigation, this paper trail will prove invaluable to the defense from the perspective of records and

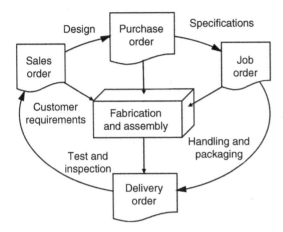

Figure 4.2 A paper trail of manufacturing. *Source:* From The Role of Sarbanes-Oxley and ISO 9001 in Corporate Management: A Plan for Integration of Governance and Operations © 2012 William A. Stimson by permission of McFarland & Company, Inc., Jefferson, NC. www.mcfarlandpub.com.

controls. Documentation reduces the risk of liability by reflecting due diligence and standard of care.

Of course, the documented system must be properly implemented. James Kolka, an internationally known attorney specializing in standards compliance, writes that ISO 9001 is a double-edged sword in litigation. It could be used by the defense in the event of a lawsuit, but the outcome depends upon the quality of the system and the quality of the paper trail. He writes, "The good news about ISO 9000 is that you have a paper trail in the event of a lawsuit and the bad news about ISO 9000 is that you have a paper trail in event of a lawsuit" (Kolka, 2004).

If a performer has a malfunctioning quality system traceable to management misfeasance, the prosecution can use such evidence against the defense. Conversely, if the paper trail reveals good business practices and internal controls, the defense can use such evidence to demonstrate prudent performance.

Quality and reliability of a product are statistical properties of the processes that provide them. The probability of failure of any product, no matter how good the process, is not zero. It is always possible for a product, one of many, to fail. Producers know this and so do judges. A "rare event" plea can carry weight in a single failure. Even when great damage is done, negligence may be difficult to prove. But if the operations that made the failed product are at fault, there may well be systemic failure. This means the performer cannot measure the quality of its products and strict liability or negligence can be more easily established.

Hence, the paper trail will reveal a performer's state of affairs one way or another. A partial list of the documentation in a paper trail of business is shown in Table 4.2, in order to get a flavor of issues that might seem bureaucratic

Table 4.2 Partial list of documents in conduct of operations.

• Authorizations	• Measurements
• Change orders	• Policies and procedures
• Contracts	• Purchase orders
• Criteria and tolerances	• Retention periods
• Delivery orders	• Signatures
• Dispositions	• Test and inspection results
• Job orders	• Traceability

to the efficiency expert, but that may be absolutely necessary to the forensic investigator.

Following the collapse of customer confidence in the aftermath of corporation scandals in 2001, the US Department of Justice became very interested in paper trails and in controls. In law, they are not form but substance, and a company can get into trouble if the trail is not clear. In forensics, both evidence and the absence of evidence may be of interest. An absence of evidence suggests misfeasance at best and perhaps malfeasance. For example, under Title XI of the Sarbanes–Oxley Act, if certain records sought by the plaintiff are not available, an investigation into this absence will be initiated, with severe penalties if the reason for absence was to avoid the law (United States Congress, 2002).

4.2 Form and Substance

A forensic systems analyst—indeed, any analyst—requires a model from which to make comparisons of what is to what should be. Because it is universally recognized, I have chosen ISO 9001 for the model of performance management to be used in this book, but it is an uneasy choice. The reason for the unease is that ISO 9001 changes form so frequently that almost anything invoking ISO 9001 today will be, as a reference, out of date tomorrow. However, although the standard frequently changes form, it does not change substance and it is the latter that the forensic analyst must pursue. So let us distinguish the difference.

"It's all form and no substance!" We have all heard this claim at one time or another. The speaker usually means that the thing being talked about is all show and no action. Since we understand what is meant, there is a common sense that form and substance are distinct. In everyday speech we tend to agree that form refers to how a thing looks; substance refers to what a thing does.

In regard to *form*, the Dictionary Britannica offers a number of definitions, but they all boil down to the intelligible structure of a thing that is distinguished from its matter. This idea suggests a very simple working definition that we shall adopt for this book: *Form is a model.* A model represents something to be made or is already existing, or once existed. We might consider the US Constitution as a model of government.

In regard to *substance*, the same dictionary provides three distinct definitions: (i) the material of which any thing is made; (ii) that which gives stability or confidence; and (iii) the essential nature in which qualities or attributes adhere. The first definition says that a substance is made of material, thereby implying that all substance is matter. However, the second and third definitions do not require matter. This is an important distinction for something without matter, say, knowledge and character, fits the other definitions.

We do not need to resolve this dichotomy because for our purposes, definitions (ii) and (iii) suffice. Substance is the essential nature of a thing in which qualities or attributes adhere and which gives stability or confidence. For example, we might consider the Declaration of Independence as the substance of our system of governance, for which the Constitution is the model. In principle, the two are tied together: the Declaration providing the substance, the Constitution providing the form.

But what are substantive business processes and how do you identify them? How do you even find them in a standard such as ISO 9001, which changes form regularly? For example, we might decide that documentation is substantive, but its requirements may appear in Clause 3 of one version of ISO 9001, in Clause 4 in another, and may not be listed at all in yet a third. Consider that product reliability appears nowhere in any version of ISO 9001, but it is implied in various different clauses. The requirement for a given process may be quite firm but implicit. Forensic analysts must be both sleuths and strongly formed in the English language as well.

4.2.1 Reference Performance Standards

A method that I use to determine substantive controls is to examine several international and national standards of performance and then adopt those controls held in common as universal good business practices. Three such influential standards are offered by the Committee of Sponsoring Organizations of the Treadway Commission (COSO, 2013); the Information Technology (IT) Governance Institute (2007), and the International Organization for Standardization (ISO, 2015). The framework of control organization used by the IT Governance Institute is discussed in some detail in Chapter 5. It is quite similar to that of ISO 9001 shown in Figure 4.1 in that it indicates the general nature of effective process control achieved through a feedback loop of verification and validation.

The IT framework contains 34 internal controls in four categories. It is beyond our scope to list them all, but for purposes of demonstrating commonality

Table 4.3 IT monitoring factors mapped to ISO 9001.

IT monitoring factor	ISO 9001 (ANSI/ISO/ASQ, 2015) control (clause number)
M1: Monitor the processes	9.0
M2: Assess internal control	4.0, 8.0
M3: Ensure compliance	4.0, 5.0, 8.0, 9.0
M4: Provide governance	4.0, 5.0, 6.0, 9.0

between standards, we shall focus on the controls in *Monitor and Evaluate*, which are M1: Monitor the processes; M2: Assess internal control; M3: Ensure compliance with external requirements; and M4: Provide IT governance. Table 4.3 maps these controls to those of ISO 9001, demonstrating close correlation.

COSO also uses a framework similar to that of Figure 4.1 in its control structure, assigning controls in two categories: the corporate environment and the corporate activity. Control environment refers to those controls that provide direction and establish degrees of operational freedom. This is traditionally top management responsibility.

The control activity refers to those controls that get things done. This activity spans the company, vertically and horizontally. All levels of the company hierarchy and all of its processes are engaged in control activities. Process and product assessment, and verification and validation are an important part of the corporate activity. All of the COSO controls, both environmental and activity, map perfectly into those of ISO 9001:2008 (Stimson, 2012).

4.2.2 Forensics and the Paper Trail

A standard is, by definition, an agreement among participants to do something in a certain way. Hence, it is voluntary. This means that the basis of every standard is compromise, because if a criterion is seen as too difficult or too cumbersome, then you have few participants and no standard. So the final document will be a careful balance between what the experts believe the standard ought to say and what the participants will accept. This dichotomy is at the root of all dissension with respect to ISO 9001. Some believe that it is too tough; others that it is not tough enough. Every revision of the standard involves another battle over the requirements.

A case in point was mentioned earlier, having to do with product reliability, which has a quite rigorous definition, given in Appendix B. For the moment, let us simply say that product reliability is quality over time—how long will the product last? You would think that producers would vaunt product reliability in their sales, but they rarely do, because adding reliability adds cost. Thus, no version of ISO 9001 has ever required product reliability and this absence is not by accident. Too many performers simply do not want it. Yet, you can find the requirement for reliability if you look hard enough. The writers of standards know English too and, indeed, they are quite good at it.

One of the irritating issues in past ISO 9001 versions has been its insistence on documentation, which many performers find are overwhelming. I imagine that every standards meeting includes a battle over which documentation requirements can be deleted and which must be retained. The issue is critical to forensics of any kind, which is fundamentally a review of what happened. If there is no record, there can be no review. I discussed this issue earlier in reference to Figure 4.2, which indicates at least a minimum set of documentation requirements.

However, ISO 9001:2015 vaunts a great reduction in documentation. Must forensic systems analysts simply throw up their hands in surrender? No, because the 2015 version of ISO 9001 is strong in risk management. So you attack the problem from another direction. If there is no explicit requirement for say, a test, but its absence increases consumer risk, then the requirement for testing is implicit and defendable in litigation.

Credits

This chapter consists of various sections taken from *The Role of Sarbanes-Oxley and ISO 9001 in Corporate Management: A Plan for Integration of Governance and Operations* © 2012 William A. Stimson by permission of McFarland & Company, Inc., Jefferson, NC. www.mcfarlandpub.com.

References

ANSI/ISO/ASQ (2005). *ANSI/ISO/ASQ Q9000-2005: Quality Management Systems—Fundamentals and Vocabulary*. Milwaukee, WI: American National Standards Institute (ANSI) and the American Society for Quality (ASQ).

ANSI/ISO/ASQ (2009). *ANSI/ISO/ASQ Q9004-2009: Quality Management Systems—Managing for Sustained Success of an Organization*. Milwaukee, WI: ANSI and ASQ.

ANSI/ISO/ASQ (2015). *ANSI/ISO/ASQ Q9001-2015: Quality Management Systems—Requirements*. Milwaukee, WI: ANSI and ASQ.

Committee of Sponsoring Organizations of the Treadway Commission. (2013). *Internal Control—Integrated Framework Executive Summary*, p. 3. https://www.coso.org/Documents/990025P-Executive-Summary-final-may20.pdf. Accessed September 19, 2017.

International Organization for Standardization (ISO) (2015). Geneva, Switzerland: ISO. http://www.iso.org. Accessed November 13, 2015.

International Organization for Standardization (ISO) (2016). Geneva, Switzerland: ISO. http://www.iso.org. Accessed November 13, 2016.

IT Governance Institute (2007). *Control Objectives for Information and Related Technology 4.1.* Rolling Meadows, IL: IT Governance Institute, pp. 29–168.

Kolka, J. W. (1998). *ISO 9000: A Legal Perspective.* Milwaukee, WI: ASQ Quality Press, p. 61.

Kolka, J. W. (2004). ISO 9000 and Legal Liability. White paper to clients, January 2004.

MIL-Q-45208 (1963). *Military Specification: Inspection System Requirements.* Washington, DC: United States Department of Defense.

MIL-Q-9858A (1993). *Military Specification: Quality Program Requirements.* Washington, DC: United States Department of Defense.

Shewhart, W. A. (1931). *Economic Control of Quality of Manufactured Product.* Princeton, NJ: Van Nostrand.

Stimson, W. A. (2012). *The Role of Sarbanes-Oxley and ISO 9001 in Corporate Management.* Jefferson, NC: McFarland & Company, Inc., pp. 102–103.

United States Congress (2002). *H. R. 3763: The Sarbanes–Oxley Act of 2002, Title XI: Corporate Fraud and Accountability.* Washington, DC: United States Congress.

West, J., Tsiakais, J., and Cianfrani, C. (2000). "Standards Outlook: The Big Picture." *Quality Progress*, January, pp. 106–110.

5

The Materiality of Operations

In this book, "operations" refers to the productive and service processes used in business and "quality" refers to the degree of conformity to which the resulting products and services meet customer requirements. Generally, the connection between the processes and the goodness of delivered product or service is hard to make because many of the processes occur early on in the system. As a result, the contribution to product quality made by the various processes in the stream of activity is often misunderstood.

As an example, suppose that a printer has been using a perfectly good three-color printer that has satisfied the customer base for years. Then a new customer arrives who insists on four colors. The printer does not want to lose this customer and buys a new four-color printer. This is usually considered a cost of production. It is not. It is a cost of quality. It is the cost of meeting customer requirements—the fundamental task of quality.

Hendricks and Singhal (1999) explain the financial advantage that quality brings to those who practice it diligently, as opposed to those who don't. However, I believe the case for quality is not obvious at all to the top floor and there are many reasons for that. Nevertheless, an effective quality management

Forensic Systems Engineering: Evaluating Operations by Discovery,
First Edition. William A. Stimson.
© 2018 John Wiley & Sons, Inc. Published 2018 by John Wiley & Sons, Inc.

system (QMS) requires that top management control all their corporate finances, including those of quality.

There are two aspects to measuring financial control: materiality and liability. When quality is expressed in these terms, then its value is clear to executive management. In this chapter, we want to measure the cost of quality in materiality and liability because they are regarded as critical financial measures of any company, hence fundamental to operations and to litigation. Managers must understand the relation of cost of quality to operations in order to make wise decisions. Attorneys must understand this relationship in order to plan an effective litigation strategy.

5.1 Rationale for Financial Metrics

In the era of mass production, the quality of product is no longer assured. Instead, the notions of producer and consumer risks are adopted and product or service quality becomes problematic. Top management determines an acceptable producer's risk and compensates for the increased consumer risk through warranty or marketing ploys. Why would top management choose lesser quality if this option were really poor economic strategy? The answer is that the cost of quality is expressed by a large variety of metrics, many of them obscure to top management.

Reporting the cost of quality in terms of materiality attracts the attention of management because these are the metrics they learn in business schools. Moreover, materiality reveals the true weight of quality on operations and on liability.

There are two aspects of the cost of quality: the cost of conformance and the cost of nonconformance. The former are all those costs required to make things right the first time, from competent management and operators to top-of-the-line equipment. Expressed as strategic goals of the company, the cost of quality is a critical appraisal and one measure of how well the company does what it does.

5.1.1 Sarbanes–Oxley

Although the Sarbanes–Oxley Act (United States Congress, 2002) is not in general germane to forensic systems engineering, certain ideas intrinsic to the Act are important to analysis. SOX is divided into 11 titles that mandate strict requirements for the financial accounting of public companies. Only two of them are needed to develop a rationale for financial metrics in quality: Title III: Corporate responsibility and Title IV: Enhanced financial disclosures.

5.1.1.1 Title III: Corporate Responsibility
Section 302 requires the CEO and CFO of every company filing periodic financial reports under the Securities Exchange Act of 1934 to certify that

the reports are true and fairly present the true financial condition and results of operations. This protects the investor. It concerns forensic systems analysts because if the cost of quality is material, then it must be accurately reported. There is only one way to know if the cost of quality is material and that is to measure its true cost in financial terms. The true cost of quality is discussed later in this chapter in the section on mapping the cost of operations to finance.

5.1.1.2   Title IV: Enhanced Financial Disclosures

Section 404 holds that top management is responsible to verify the effectiveness of internal controls in finance and in operations. What is not yet resolved is the extent of operations that must be verified. At this time, the consensus is that information technology (IT) controls must be verified because much financial data resides in the IT system. Beyond this, Section 401 is ambiguous in its use of the word, "operations," the nature of which is not specified. However, finance is just an account of doing business. There is a cost to operations and it is prudent to assume that financial disclosure refers to all corporate operations, not simply those of the accounting department. The prudent CEO will anticipate a requirement for effective controls in any activity that could affect the market value of the company.

5.1.2   Internal Control

The definition of internal control used by the SEC comes from the Committee of Sponsoring Organizations of the Treadway Commission (COSO, 2013): An internal control is a process designed to provide reasonable assurance regarding the effectiveness of operations, reliable records and reports, and compliance with regulations. This definition is practical because, correctly implemented, it is process oriented and conforms to traditional engineering notions of system effectiveness and stability.

"Internal control" implies a controlling structure as well as a process to be controlled. As an example, The IT Governance Institute lays out a system of performance as shown in Figure 5.1, with four kinds of processes and 34 controls. Each process addresses a subset of the 34 controls provided by the ITGI standard "Control Objectives for Information Technology" (CobIT, 2000). Figure 5.1 is a model for IT operations controls and is quite similar to that of ISO 9001, as shown earlier in Chapter 4. Each is a business process model with various activities assigned to each subprocess.

The four major processes are plan and organize; acquire and implement; deliver and support; and monitor and evaluate. The controls of planning and organizing for CobIT are listed in Table 5.1, in the left hand column. These controls concern management performance and are strikingly similar to some of the requirements of Clauses 4–9 of ISO 9001 (ANSI/ISO/ASQ, 2015) shown in the adjacent column.

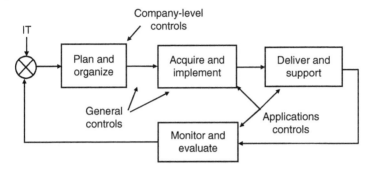

Figure 5.1 The IT Governance Institute model of performance. *Source:* Stimson and Dlugopolski (2007). Reproduced with permission from *Quality Progress* © 2007 ASQ, www. asq.org.

Table 5.1 Comparison of some controls for CobIT and for ISO 9001.

CobIT list of internal controls for planning and organization	ISO 9001 requirements, Clause 4
Define strategic IT plan	Identify processes for QMS
Define information architecture	Establish documentation system
Define technological direction	Ensure availability of resources
Define IT organization and relationships	Define organization and relationships
Manage the IT investment	Define strategic quality plan
Communicate IT aims and direction	Communicate aims and direction
Manage human resources	Manage human resources
Ensure compliance to external requirements	Ensure compliance to external requirements
Assess risks	Assess risks
Manage projects	
Manage quality	

Source: Stimson (2007). Reproduced with permission from Quality Progress © 2007 ASQ, www. asq.org.

Similar comparisons can be made from each of the CobIT controls shown in Figure 5.1 to the requirements of various clauses in ISO 9001. The law recognizes that the CobIT requirements satisfy the SOX criteria for internal controls. As the ISO 9001 requirements are comparable to those of CobIT, they can therefore be regarded as internal controls for quality of product and service.

5.1.3   The Materiality of Quality

The International Accounting Standards Board (IASB, 2001) defines materiality as follows: "Information is material if its omission or misstatement could influence the economic decisions of users taken on the basis of financial statements."

Materiality is a financial issue. How does this affect quality? Financial statements are about the costs of doing business. They include the costs associated with quality, which is the degree to which processes are effective. When the cost of quality is expressed in terms of the general ledger, its materiality is transparent and may be important. If the cost of quality is not identified, it is an omission in the financial statement and if it is large enough, then it becomes material and its omission is a Federal crime.

Materiality is not a science. A rule of thumb says that if the financial error is greater than 5%, the cost is material. But 5% of what? References may be net income, gross profit, total assets, total revenue, or percentage of equity. The law is not settled on this and eventually a convention will have to be agreed upon.

Materiality has never been an issue with quality, so why now? There are several reasons. We have already discussed the first, that the cost of quality has long been neglected simply because it is so poorly understood in the executive suite. The second reason is the new Sarbanes–Oxley law, which criminalizes gross errors in materiality. The third reason has to do with an increasing ethical awareness in industry. Judy Kuszewski (2004) points out that corporate responsibility is an increasing part of annual financial reports and that these issues go well beyond the bottom line—they get to the heart of the business, the impact on people, the environment, and company values. In short, the law is becoming increasingly aware of corporate social responsibility.

5.2   Mapping Operations to Finance

There is a continuum of financial events and of operations in the corporate life cycle, keyed to each other and the task is to match them up—to balance the books. This task may be tedious, but it is not hard. You go through a list of quality issues and simply match each element to a factor in the general ledger, which tracks assets, liabilities, incomes, and expenses. It tells you the current year transactions and beginning and ending balances and net income statement and is a key document for audit trails.

The general ledger is initialized every year, which means that some accounts are still outstanding—purchases paid for but not yet delivered; purchases delivered but not yet paid for, and so on. Even without integration, the accounting trail is tough because it extends over several ledgers, but if other departments are brought into consideration, the task may become easier. A strategic picture would be absolutely required in which each activity is agreed on the assignment of a cost. This would reduce redundancy, prevent omissions, identify the status of each purchase and payment, and simplify and clarify the cost accounting.

For example, you receive a large nonconforming shipment, partially paid for in advance. Only the end user is aware of the nonconformity and will insist that its status be identified. This will force Purchasing, Production, Quality, Finance, and Tax to agree to a convention on how such an item will be accounted for. The system of governance will eventually streamline to easy-to-audit uniform policies, procedures, and accounting.

To match operations to finance, the costs of quality must be expressed in terms of the general ledger. Many of these costs occur as a result of poor quality, but poor quality is only one part of the cost of quality and in a very good company, may be a very small part. Capable managers, operators, and equipment are expensive, so strategic quality is not free. The left column of Table 5.2 presents a partial list of the costs of quality. You could easily think up more, and that is the problem. Not only does this vast array confuse most CEOs, but it scatters the effect. Many small costs are not necessarily an obvious big cost. The many measures of quality must be mapped to just a few items in the general ledger, to focus their true cost.

This mapping is shown in the right column of Table 5.2, which contains an arbitrary convention. Any cost incurred through operations is an operational cost—ipso facto. This is not as informative as it sounds because some costs of operations are due to the expense of running the show, which is necessary, and others are due to poor quality, which is not necessary. So we distinguish between the expense of operations, which you must have to stay in business, and the cost of operations due to unnecessary work.

The reason to distinguish costs from expenses is that net income is proportional to net sales less operating costs. Too many executives take the chainsaw approach—"Aha! To get more net income, we have to cut operating costs! Start firing people and sell off stuff till we get in the black again!" But if they amputate their sources of effectiveness they never will.

Losses are such things as canceled orders and unpaid invoices that are lost because of poor quality. They are the negative of accounts receivable. Accounts receivable is money coming in. Loss is money *not* coming in. Total assets are proportional to accounts receivable, so a lesser account reduces cash flow.

The point of all this is not so much to reduce cost of quality as it is to show executive management that the cost of quality is material to profitability and to the market value of the company. It is within the purview of the law. Forced to consider cost of quality, the prudent CEO will see how quality can affect the bottom line *and* contract compliance.

It may be hard to imagine that the quality of product or service is material to the value of a major corporation with a good product line and millions of dollars in assets. But critics claim otherwise. Kaner (1996) estimates the cost of quality

Table 5.2 The costs of quality.

Category	Measures of quality (partial list)	Mapping of quality costs (item of the general ledger)
Failure	Scrap; rework; labor; sorting; downtime; slowdowns; complaints; investigations; travel; recall; unpaid invoices; lost sales	Operating costs Operating expenses (labor) Variable expenses Losses
Appraisal	Receiving; in-process and final inspection; test equipment; test technicians; special tests; lab maintenance; quality control; QC overhead	Operating expenses Fixed expenses Depreciated assets (equipment) Fixed assets (technicians)
Prevention	Risk management; quality planning; design tolerances; training; housekeeping; special sourcing; packaging; life cycle tests; field tests; preproduction tests; shelf tests; cash flow; inventories	Operating expenses Fixed expenses Variable assets (cash flow) Inventory

Source: Stimson (2007). Reproduced with permission from Quality Progress © 2007 ASQ, www.asq.org.

as varying from 20 to 40% of sales. In different terms, Mills (2006) estimates the cost of quality as varying from 15 to 30% of operating costs. Still a third estimate is offered by the Eagle Group (2006), claiming the cost of quality varies between 25 and 40% of revenues. These estimates indicate that the cost of quality is material because their magnitude could indeed influence financial statements. Correct accounting considers both the cost of poor quality, as shown in the "failure" row of Table 5.2, and the expense of good quality, much of it usually thought of as the cost of production. In Chapter 16, we shall see that in many attempts to reduce the cost of production, the victim is often unit quality.

5.2.1 The Liability of Quality

Every company has an average operational cost of quality. Average means what it says—about half the time the operational cost will be below average, half the time above and perhaps even into the range of materiality. You cannot know without tracking it. The cost of quality may even drift into the range of probable liability. All of us as individuals must estimate our own personal liability when we buy home insurance and auto insurance. So also, a company must be realistic about its potential liability and whether this liability can be affected by the corporate cost of quality.

5.2.2 The Forensic View

It is one thing to find evidence of system misfeasance. (Malfeasance is also a concern, of course, but this judgment will be made by an attorney.) It is quite another to determine liability. Yet, in cases such as false claims, where liability is a central issue, the costs of poor operations must be determined and the forensic systems analyst may be able to make a significant contribution in the assessment.

In the global economy, companies are using activity-based accounting more often. Describing financial analysis in terms of processes ties financial data to specific operations. By identifying nonconforming controls in operations and by mapping the estimated costs of subsequent losses to the general ledger, the forensic team can arrive at a reasonable estimate of liability.

Credits

This chapter is in large part taken from "Financial Control and Quality." W. Stimson and T. Dlugopolski (2007). *Quality Progress*, vol. 40, no. 5, pp. 26–31. Reprinted with permission from *Quality Progress* © 2007 ASQ, www.asq.org.

References

ANSI/ISO/ASQ (2015). *ANSI/ISO/ASQ Q9001-2015: Quality Management Systems—Requirements*. Milwaukee, WI: ASQ Quality Press.

CobIT (2000). *Framework of Control Objectives for Information Technology (CobIT)* ®. Rolling Meadows, IL: IT Governance Institute and the Information Systems Audit Foundation.

Committee of Sponsoring Organizations of the Treadway Commission (2013). *Internal Control—Integrated Framework Executive Summary*, p. 3. https://www.coso.org/Documents/990025P-Executive-Summary-final-may20.pdf. Accessed September 19, 2017.

Eagle Group (2006). *Cost of Quality Workshop*. www.eaglegroupusa.com. Accessed September 1, 2017.

Hendricks, K. and Singhal, V. (1999). "Don't Count TQM Out." *Quality Progress*, April, pp. 35–42.

International Accounting Standard Board (2001). *Framework for the Preparation and Presentation of Financial Statements*. https://www.iasplus.com/en/resources/ifrsf/iasb-ifrs-ic/iasb. Accessed September 14, 2017.

Kaner, C. (1996). *Quality Costs Analysis: Benefits and Risks*. Copyright Cem Kaner. http://www.kaner.com/pdfs/Quality_Cost_Analysis.pdf. Accessed September 14, 2017.

Kuszewski, J. (2004). "Materiality." Brief of the *SustainAbility Consultancy*, December 23.

Mills, D. (2006). "Cost of Quality." *iSixSigma LLC*. www.iSixSigma.com. Accessed September 1, 2017.

Stimson, W. and Dlugopolski, T. (2007). "Financial Control and Quality." *Quality Progress*, vol. 40, no. 5, pp. 26–31.

United States Congress (2002). *H. R. 3763: The Sarbanes-Oxley Act*. Washington, DC: United States Congress.

6

Process Liability

Process liability is a new concept, built upon the notion of product liability, the latter long recognized in law. The two must be distinguished, so I begin at the beginning: what do we mean by the term "liability?" In its simplest terms, a liability is a debt or obligation. The Britannica Dictionary defines liability as "the condition of being responsible for a possible or actual loss, penalty, evil, expense, or burden." If you are liable, you owe or will owe somebody something. Parents are liable for the welfare of their children. In accordance with our Constitution, our government is liable for the general welfare. (I make no attempt here to define what that means.) Performers are liable for the results of what they do.

In law, liability refers to legal or financial responsibility and applies in both civil and criminal law. If a provider is liable for injury or harm to someone, the injured party is entitled to damages, usually monetary compensation. Under tort law, legal injuries include emotional or economic harm to reputation, violations of privacy, property, and constitutional rights as well as physical injuries.

Product and service liability, therefore, refers to the responsibility for damages that a provider owes to a customer or user who is injured in some way by the said products or services. Damages refer to the sum of money that the injured party may be declared entitled to by a court of law. Actual damages are awarded in order to compensate the injured party for loss or injury. Punitive damages are awarded to punish the responsible party. In some cases, treble

Forensic Systems Engineering: Evaluating Operations by Discovery,
First Edition. William A. Stimson.
© 2018 John Wiley & Sons, Inc. Published 2018 by John Wiley & Sons, Inc.

damages are awarded—three times the amount of actual damages—when certain conditions of culpability are found.

Most of us have a fair notion of product liability; for example, there is the case where a person was injured by a tool that had been purchased and he subsequently sued the tool maker for injury (*Greenman v. Yuba Power Products, Inc.*, 1963). An example of service liability is seen in the well-known "hot coffee" case, in which a customer ordered coffee at a McDonald's restaurant and was severely burned when she spilled the coffee in her lap (*Liebeck, Stella v. McDonald's Restaurants, Inc.*, 1994).

The law of products liability governs the private *litigation* of product accidents. Operating *ex post*, after a product accident has already occurred, its rules define the legal responsibility of providers for the resulting damages (Owen et al., 2007a).

The intent of the liable party may be irrelevant. A provider operating within the law and in accordance with the standards of industry is nevertheless liable for injury caused by its products or services. Under the theory of strict liability, providers may be found criminally or civilly liable without the need to establish intent. Strict liability provides that a person may be held liable for acts regardless of whether the acts were committed with intent or through negligence or by accident, inadvertently, or in spite of good faith efforts to avoid a violation (Spencer & Sims, 1995).

A review of the literature shows that systemic failure is a significant source of liability suits (Owen et al., 2007b). Malfeasance in manufacturing operations occurs even in the presence of major performance initiatives as ISO 9001 and Lean Six Sigma, implying that standards are no protection against liability if they are not properly implemented and maintained.

Productivity is generally recognized in industry as a key indicator of the state of operations. The concept is taught in all graduate schools of business and high-level managers are familiar with it. But on the factory floor productivity means something else entirely. It is not a key indicator—it is a mandate. Many supervisors and lower level managers, who may have little understanding of product liability or of risk management, have one overriding objective—get the product out the door. This unwritten mandate is the seed of misfeasance, systemic failure, and process liability.

6.1 Theory of Process Liability

The theory of process liability was first expressed by attorney Leonard A. Miller (1970) in the *New England Law Review*. This seminal paper was entitled "Air Pollution Control: An Introduction to Process Liability and Other Private Actions." The author introduced the concept of the liability of processes and traced legal precedents that justified its use. Although framed in environmental pollution control, Miller's arguments were expressed in general terms so their applicability to general business operations is straightforward and provides a solid grounding for this chapter. Key extractions from his paper appear in the following paragraphs (with permission: *New England Law Review,* vol. 5, p. 163 (1970)). Statements in parentheses are mine and are meant to clarify a sentence in which some deletions have been made for brevity.

Through vigorous legal action in the legislature and the courts, an industrial society can achieve responsible actions in the *process* of production.

Looking at manufacturing, it can be said that generally and simply one starts with raw material, capital and labor and produces an end product. But one also produces pollution. Indeed, for every product a certain amount of pollutant is an integral part. The amount of pollution is determined by the process of manufacturing.... In a sense then, the process becomes the product. The process is determinative of the extent of the pollution product. It is this relationship which is at the base of the theory of process liability. If a process is a product, then as the manufacturer is liable for the reasonably foreseeable consequences of the usage of his product, so should he be responsible for the consequences of the process so employed. If he creates pollutants and they injure, he should be responsible, and his process of manufacturing should be at issue.

Applying these legislative statements (concerning the environmental costs and degradation due to industrial processes) and our observations about the product/process relationship, we feel that courts may wish to place more emphasis on the process of manufacturing during the course of a tort suit. We see at least three ways in which the process of manufacturing could influence the outcome or resolution of a tort suit: a) as an element in the balancing of the equities; b) as a form of relief; and c) as a new cause of action. These (judicial) uses of the process of manufacturing, and any others that might be developed, are what we choose to call process liability.

The first usage of process liability might be an element in the balancing of the equities...(in which) the courts often resort to a judicial weighing of interests. The second use of process liability is therefore to directly affect the determination of a method of production. A change in the method of production may be a part of the request for relief. The third use of process liability is the only one which could be called a new "liability" in strict legal terms. Process liability could become a cause of action related to product liability.

Process liability, under the applications of product liability, would not look to the product to find a basis for recovery, but rather to look to the process employed in manufacturing the product. In all other respects, process liability would echo product liability. This is possible since the rationale between the two types of liability is similar. Product liability makes a manufacturer responsible for the reasonably foreseeable usages of his product so that the manufacturer will take care in producing the product and so that the damages from a faulty product will be on the one responsible, i.e., the manufacturer. Process liability would make a manufacturer responsible for the reasonably foreseeable usages of a particular method of production, so that damages to individuals or the environment from that method of production would rest on the one who chose the method, i.e., the manufacturer.

In process liability there should also be a limit to the liability and this limit should be based on the concept of fault. If the defendant has shown that he has utilized the best available technology, he should not be liable in process liability. The purpose of the doctrine of process liability is to promote the usage of the process of manufacturing which would least pollute the environment. Therefore, liability should not be imposed where an industry uses the best available methods of controlling pollution. This limitation on process liability does not mean that an industry would not be liable for another tort merely because it was not liable for process liability.

6.1.1   Operations and Process Liability

Miller's theory of process liability, framed within an environmental argument, can easily be shown applicable to the manufacture of nonconforming units by simply changing analogous terms. For example, one could say, "The process is determinative of the extent of nonconformance of the product"; and "If the producers create nonconforming products and they injure, they should be responsible and their process of manufacturing should be at issue." Miller's term "method of production," too, has an equivalent meaning to "good business practices," because the method would be the process used in the provision of units. Similarly, the theory is applicable to service operations also.

The agreement of Miller's argument to general manufacturing is again displayed in his rationale for exempting process liability where the "best available methods" are used, although the defendant might still be liable for another tort. This reasoning is in accord with the engineering acceptance that no production system is perfect and that there will always be nonconforming product, which production should, nevertheless, be a rare event. The "rare event" defense against liability will be discussed in a later chapter, along with the statistical basis behind it.

Process liability is almost always the result of a working environment of misfeasance and the concept is straightforward. Management misfeasance leads to systemic failure. Systemic failure leads to product liability on a very large scale, from which process liability follows. The production process is unstable and its failure rate is indeterminate.

The probability of failure of any product or service, no matter how well performed, is not zero. It is always possible for a product or service, one of many, to fail. If there is only one failure, liability may be difficult to establish even when great damage is done. But if the operations that made the failed product or performed the failed service are at fault, there may be systemic failure. This means the producer has no idea of the quality of its performance and strict liability or negligence is more easily established. The result can be massive liability costs. Systemic process failure can result in great liability in two ways: a customer buys in large volume and sues for redress or many customers buy in smaller volume; however, they then band together in class action.

With increasing frequency, businesses today find themselves facing huge damages from liability litigation. Two relatively recent examples are the Ford Firestone litigation of 2001 (Daniels Fund Ethics Initiative, 2012) and the Toyota litigation of 2010 (Bensinger & Vartabedian, 2011). Insurance companies, who may cover liability costs, have growing concern about the responsibility of insured businesses to the quality and reliability of their products and services (Boehm & Ulmer, 2008a). Apart from the actual costs of injury, liability awards may be doubled or tripled if it can be shown that the performers were negligent or derelict in their duties.

6.1.2 Process Liability and Misfeasance

In operations, systemic process failure implies management negligence. I said earlier that W. Edwards Deming (1991) estimates that 94% of system problems are the responsibility of management. Joseph M. Juran (1992) puts the figure at 80% but either estimate is unacceptable. Misfeasance occurs when good manufacturing practices are abandoned or deviated from and may occur in any phase of performance, from design and fabrication through testing and delivery. If managerial negligence can be established, the risk of liability is greatly increased as trial courts are showing increasing interest in how quality assurance procedures are managed (Boehm and Ulmer, 2008b).

A few examples in process liability derived from misfeasance in operations are failure to comply with specifications, inadequate process inspection, and omission of reliability in process or product safety. Misfeasance invariably leads to systemic process failure, which in turn leads to degraded product reliability. Hence, product unreliability is a key indicator of systemic failure and invites inquiry into management of operations.

Misfeasance refers to improper performance as distinguished from *malfeasance*, which is illegal performance. However, misfeasance can lead to malfeasance, as for example when there is such intense focus on meeting production quotas that proper procedures are abandoned. This kind of scenario is described in Section 7.1.3.1.

6.2   Process Liability and the Law

Historically and currently, customer injury or harm is viewed within a context of product liability. A claim of injury caused by a product is made and the existence of a defective product is necessary to that complaint. This position is held even if there are hundreds of injuries. Then hundreds of bad products must be found. Process liability is, at present, just an idea. The possibility that a nonconforming system may be producing hundreds or thousands of nonconforming units unknowingly, because no one is looking, is still beyond judicial consideration.

Although the idea of process liability is not yet explicit in the judicial system, courts are moving in that direction in that they are increasingly concerned with management use of good business practices. Chapter 15 of this book discusses the cause and effect relation between process and product, which relation is not yet established in law, but we are getting there. Decisions in class action suits indicate that the law now views service and manufacturing as a set of interrelated processes. A few cases in which the trend to process liability is evident are discussed in Chapter 8.

Credits

This chapter consists of several sections taken from *The Role of Sarbanes-Oxley and ISO 9001 in Corporate Management: A Plan for Integration of Governance and Operations* © 2012 William A. Stimson by permission of McFarland & Company, Inc., Jefferson, NC. www.mcfarlandpub.com. This chapter contains copyrighted material in the form of direct quotes from Leonard A. Miller (1970). "Air Pollution Control: An Introduction to Process Liability and Other Private Actions." *New England Law Review*, vol. 5, pp. 163–182, with permission of the copyright holder, *New England Law Review*, Boston, MA. Permission granted December 31, 2016. Citations should be to Volume 5 of the *New England Law Review* (Blue Book cited as: 5 New Eng. L. Rev 163 (1970)).

References

Bensinger, K. and Vartabedian, R. (2011). "Toyota to Recall 2.17 Million More Vehicles." *Los Angeles Times*, February 25, 2011. http://articles.latimes.com/2011/feb/25/business/la-fi-toyota-recall-20110225. Accessed September 14, 2017.

Boehm, T. C. and Ulmer, J. M. (2008a). "Product Liability Beyond Loss Control—An Argument for Quality Assurance." *Quality Management Journal*, vol. 15, p. 7.

Boehm and Ulmer (2008b), p. 11.

Daniels Fund Ethics Initiative (2012). Firestone's Tire Recall. Anderson School of Management, University of New Mexico. http://danielsethics.mgt.unm.edu. Accessed October 24, 2012.

Deming, W. E. (1991). *Out of the Crisis*. Cambridge, MA: Massachusetts Institute of Technology, p. 315.

Greenman v. Yuba Power Products, Inc. (1963). 59 Cal.2d 57. http://scocal. stanford.edu/opinion/greenman-v-yuba-power-products-inc-27186. Accessed September 19, 20175.

Juran, J. M. (1992). *Juran on Quality by Design*. New York: Free Press, p. 428.

Liebeck, Stella v. McDonald's Restaurants, Inc. (1994). New Mexico. https:// en.wikipedia.org/wiki/Liebeck_v._McDonald%27s_Restaurants. Accessed December 24, 2015.

Miller, L. A. (1970). "Air Pollution Control: An Introduction to Process Liability and Other Private Actions." *New England Law Review*, vol. 5, pp. 163–172.

Owen, D. G., Montgomery, J. E., and Davis, M. J. (2007a). *Products Liability and Safety: Cases and Materials*, 5th ed. New York: The Foundation Press, p. 14.

Owen, D. G., et al. (2007b), p. 25.

Spencer, M. P. and Sims, R. R., eds. (1995). *Corporate Misconduct: The Legal, Societal, and Management Issues*. Westport, CT: Quorum Books, p. 49.

7

Forensic Analysis of Process Liability

In reference to business operations, process liability refers to misfeasance, malfeasance, fraud, and false claims. We usually do not think of associating fraud with operations beyond its financial aspects. Certainly, fraud happens in marketing or in accounting or perhaps in the boardroom, but in operations? However, as I commented earlier, a review of the literature shows that fraud in operations is a significant source of liability suits. If fraud were limited to its popular conception of cheating for self-gain, then such events might be quite rare. But fraud is defined much more broadly than that. There are two

Forensic Systems Engineering: Evaluating Operations by Discovery,
First Edition. William A. Stimson.
© 2018 John Wiley & Sons, Inc. Published 2018 by John Wiley & Sons, Inc.

definitions: (i) the deliberate deception practiced so as to secure unfair or unlawful gain and (ii) the representation about a material point that is intentionally false and which is believed and acted upon to the victim's harm. The first definition fits the popular image, focusing on self-benefit. The second focuses on the harm done to the victim and its wording is more easily applied to the forensic inquiry of production and service systems.

If a company accepts payment for systemic nonconforming product or service, allegations of fraud and false claims may follow. The US government and 32 states have enacted false claims acts (FCA). False claims provide for liability of treble damages for those who knowingly seek to defraud the federal or state government. Under the FCA, *knowingly* is defined as (i) has actual knowledge of the information; (ii) acts with deliberate ignorance of the truth or falsity of the information; or (iii) acts in reckless disregard of the truth or falsity of the information (United States Code 31, 2007). *Reckless* is defined in law as rash, indifferent, or neglectful. Therefore, a defendant can be cited as reckless if found responsible for indifferent or negligent management. Under false claims, executive management can be guilty of malfeasance by simply being negligent or indifferent to the tactics used by floor supervisors responding to heavy production quotas.

Is malfeasance widespread? The Corporate Fraud Task Force of the United States Department of Justice (DOJ, 2008) has used the law to obtain more than 1300 corporate fraud judgments. The DOJ list includes the dozen or so companies charged as a result of the financial scandals that followed the Enron blow up, but the remainder covers the spectrum of malfeasance. For example, from 2002 to 2010, the DOJ Civil Division (2010) had successfully litigated 65 manufacturers, of which 28 cases were settlements of fraud in production. In the same vein, the DOJ Civil Division successfully litigated cases in which the cost of operations malfeasance was over $2.15 billion. Of the successful settlements, all but one of the performers was certified under ISO 9001 at the time of the allegations, which apparently remained undetected by ISO auditors. These figures may seem small when measured against the number of businesses in the United States or against the Gross Domestic Product, but measured against a standard of ethical behavior they indicate a deplorable level of malfeasance.

The Sarbanes–Oxley Act (United States Congress, 2002) puts fresh accent on the responsibility of management by extending it to internal controls. For example, in Section 404, SOX implies its purview to operations, thereby nominally assigning to top management the responsibility for financial and operational controls. In this chapter, we focus on manufacturing operations, but the reader should understand that the analysis applies to service also. In its process structure, service does not differ from production; a service is designed; processes are developed, and the final delivered activity, while a performance, is nevertheless a product of the design and development phase of the business.

7.1 Improper Manufacturing Operations

Manufacturing operations consist of designing a product, assembling the resources necessary to its fabrication, making the product, verifying and validating the product, then selling and delivering the product to customers. Each of these steps requires good manufacturing practices. Table 7.1 lists some of the procedures that are contrary to good practices, are clearly misfeasance, and suggest the possibility of malfeasance. Not all the procedures are inherently illegal, but they invite examination because they can result in systemic product failure. In the following sections, these procedures are briefly described and classified according to the manufacturing phase in which they occur: verification and validation; resource management; and process management.

7.1.1 Verification and Validation

7.1.1.1 Nonstandard Design Procedures

This book is about systems and, in particular, the systems used in operations. Thus, verification and validation refer not just to the final product or service provided to the customer but to the processes that produce and serve, including the design process itself.

The design process is well identified, as in Arora (1989) and elsewhere. ISO 9001 (ANSI/ISO/ASQ, 2015) lists five generally recognized phases in the process of design and development: (i) planning, (ii) inputs; (iii) controls; (iv) outputs; and (v) changes. The planning phase includes customer requirements, and the design inputs include the requirements for performance and pertinent statutes. Controls include design reviews, verification and validation of the design, and documented information of these activities, such as acceptance test criteria and test results. Validation and verification control also includes the intended use requirements and implies product reliability because of the future that is implicit in the word "intended."

For a newly designed product with kinship to existing products, which is the usual case, there will be a standard design procedure either in-house or

Table 7.1 Some operations that may indicate malfeasance.

Nonstandard design procedures	Substandard purchased parts
Unverified or unvalidated design	Ghost inventory
Tests waved by management	Ineffective flow down
Altered test procedures and results	Forced production
Unmonitored outsourcing	Abuse and threats by management

industry wide. For example, reliability design will require life cycle testing, for which the mathematics and procedures are well identified. Variation from standard design procedures raise questions and invites inquiry into the integrity of the design.

7.1.1.2   Unverified or Unvalidated Design

Verification answers the question, "Are we doing things right?" Validation answers the question, "Are we doing the right things?" (Boehm, 1984). The two activities are quality controls in ISO 9001, defined in Clause 8.3.4(c), *verification* and 8.3.4(d), *validation*. The former requires verification to ensure that design and development outputs meet the input requirements. The latter requires validation to ensure that the resulting product or service meets the requirements for the specified application or intended use. Verification and validation can vary from rustic to rigorous, depending on how the producer sees its responsibilities. The more rigorous the procedures the greater the cost of production, hence there is a negative incentive to a definitive program of verification and validation. The forensic analyst is advised to look for test integrity in any program of verification or validation in litigation.

7.1.1.3   Tests Waived by Management

Testing is the means by which the quality of products is verified and validated. ISO 9001 requires a company to verify and validate the design of a product, but it also requires the same for fabrication. Clause 9.1 requires verification and validation of both the products and the processes that made them.

On occasion, a surge in demand or an obstruction in supply will cause an inadequate flow of production and create a fire drill reaction. If the pressure to increase the flow is great, management may waive testing, thereby negating the verification and validation phase. Customers must be informed of the waiver and may accept it if the alternative is delayed delivery. Unfortunately, sometimes the customer is not told that insufficient testing or perhaps no testing was done, which is both unethical and illegal if the terms of the contract required verification.

7.1.1.4   Altered Test Procedures and Results

Clause 9.1 requires verification that product requirements have been met and requires collection and analysis of appropriate data to demonstrate the effectiveness and suitability of the quality management system (QMS). These data include the results of product and process monitoring, and the measurement and analysis of data must provide information relating to the conformity of the product or service to customer requirements.

The altering of test procedures or of test results refers to changing a validated test procedure or changing the data resulting from the procedure in order to affect a particular result. Both steps violate test integrity and if the test results are knowingly false, then the alteration is malfeasance.

7.1.2 Resource Management

7.1.2.1 Unmonitored Outsourcing

In the global economy, much of manufacturing is distributed by contractors to numerous suppliers as subcontractors, which greatly increases the difficulty of supplier control. Recognizing this reality, Clause 8.1 of ISO 9001 (ANSI/ISO/ ASQ, 2015), *Operational planning and control*, requires an organization (company) to ensure that outsourced processes are controlled.

This clause is expanded upon in Clause 8.4, *Control of externally provided processes, products and services*, which appears quite comprehensive in its scope. Clause 8.4.1 holds the company responsible for all the necessary supplier controls: (a) when products and services from a supplier are to be integrated internally with the company's own products and services; (b) when products and services from a supplier are provided directly to the customer on behalf of the company; and (c) when a decision has been made by the company to outsource a process to a supplier.

Clause 8.4.2 adds that the company must ensure that (1) the provision of suppliers do not affect the conformity of products and services delivered to the customer and (2) that the processes of suppliers are controlled in conformity with their (the supplier) QMS.

Consider Clause 8.4.1(b). A subcontractor that is authorized to provide product or service directly to a customer is called a *self-release supplier*, and from Clause 8.4.2, that supplier follows its own QMS independently of the ISO 9001 certified company to whom they supply and who owns the contract. Hence, it is possible for a contract to be carried out by a prime contractor and its numerous suppliers, all having independently operating QMS, one for the prime contractor and one for each subcontractor, with little or no coordination or oversight.

Or consider Clause 8.4.1(a) and (c). In supplying parts or services to be integrated with those of the prime contractor, the supplier is responsible for the outsourced process to conform to its quality management system. The type and extent of control applied to the outsourced process is therefore defined in the said QMS. Once the control procedure is defined, some manufacturers assume that the requirement is met. They then proceed to write a supplier control procedure with as little control as they choose.

Hence the intended rigor in supplier control under ISO 9001 is often reduced to a battle of words. For example, as stated earlier, Clause 8.1 requires an organization to *ensure* control over outsourced processes, but in defending against liability, some manufacturers can and have claimed to meet this criterion with their defining procedure, even though ineffective. This ambiguity can be avoided by the ISO with the simple expedient of a revised standard with clear requirements that responsibility for product quality cannot be delegated.

Although ISO 9001 is not of much help in the matter of recalcitrance, the forensic systems analyst can rely on other authorities, Federal acquisition regulations. *FAR 52-246-2(b), Inspection of Supplies* (Federal Acquisition Regulations, 1996), admits supplier review, but it does not relieve the Contractor of the obligations under the contract. *FAR 46-405(a), Subcontracts* (Federal Acquisition Regulations, 1995), recognizes contract quality assurance on subcontracted supplies or services, but it does not relieve the prime contractor of contract responsibilities. In those cases in which Federal rules are not involved, the analyst must rely on careful reading of the requirements of the contract, the corporate management system standard in use, and any applicable industrial standards. There will almost always be a formal requirement upon the prime contractor that cannot be delegated to subcontractors. The guiding principle is quite simple: the contract is between the buyer and the prime contractor.

7.1.2.2 Substandard Purchased Parts

Products may require resources as input material and very often those resources are parts purchased from another manufacturer. The earlier versions of ISO 9001 were explicit in requiring that purchased parts meet customer requirements. However, version ISO 9001 (ANSI/ISO/ASQ, 2015) has taken a different tack by combining purchasing requirements with those of supplier control. The result is a requirement posed in negative form in Clause 8.4.2: "The organization shall ensure that externally provided processes, products and services do not adversely affect the organization's ability to consistently deliver conforming products and services to its customers."

As an aside, in my view requirements in negative form obscure rather than clarify and make the forensic task more difficult because the requirement is no longer explicit, but vague. To do something is clear; to not do something adverse is not clear because it shifts the focus on just what is adverse and what is not. In the end, the forensic analyst must determine whether a given part meets customer requirements.

Parts that do not are substandard. The deliberate use of substandard parts is malfeasance, but this is usually not the case. If supplier control is inadequate, purchased parts may be substandard because the customer requirements are misunderstood by the supplier or because the supplier procedures are ineffective. Therefore, the use of substandard parts is often misfeasance rather than malfeasance. The result to the customer is the same.

7.1.2.3 Ghost Inventory

Ghost inventory is an inventory that is claimed but does not exist, or conversely, it exists but is not claimed, all in the interest of manipulating cost accounting in one direction or the other. Ghost inventory is a good example of collusion in financial and production reports because inventory is an explicit cost of a company's general ledger. Hence, finance is aware of inventory volume, costs, and ghosts. At the same time, the existence of physical inventory or lack thereof is certainly known in the plant, so production is aware of the ghosts also.

7.1.2.4 Ineffective Flow Down

Flow down refers to the obligation that a prime contractor has to "flow down" information to its subcontractors. With the rise in outsourcing, flow down has become an increasingly significant factor in supplier control and is now a specialty in law. Common clauses that are "flowed down" include product specifications, scope of work, dispute resolution guidelines, and state and federal regulations. Some flow down clauses are mandatory; for example, in a Federal contract rules pertaining to the treatment of persons with disabilities must be flowed down from the prime contractor to all subcontractors (U.S. Code of Federal Acquisition Regulations, FAR, 2014).

Other flow down clauses may not be mandatory but are necessary for effective work. For example, again in a Federal contract, the requirements of FAR 46-405, *Subcontracts*, relative to unmonitored outsourcing also apply to flow down: the prime contractor is not relieved of any responsibilities under the contract. This idea is reinforced in an ISO 9001–based contract because of Clause 8.4.2, which requires that any external provisions do not adversely affect conformity to customer requirements.

7.1.3 Process Management

7.1.3.1 Forced Production

Usually, we describe production systems in one of two modes, either push or pull. In push production, supply governs the production rate, which tends to large in-process inventories. In pull production, demand governs the production rate and levels of inventory tend to be low or just in time (Goldratt & Cox, 1986). Although Masaaki Imai (1997) states that Japanese industry prefers pull production, both modes are thought to be useful in American business schools.

But there is yet a third mode that has no name and is not taught in business schools. I call it *forced production* for want of a better name for an unmentionable activity. Forced production is characterized by a schedule that rises to the level of desperation. Production is everything and employees are encouraged to do whatever it takes to get the product out of the door. Product tests and inspections are abandoned, waived, or altered; measurement is slovenly; overtime is mandatory; the pace of work is hurried; and employee abuse is common. Supervisors may use coercion of employees in order to meet their quotas.

In forced production the likelihood of systemic product failure is high because of the increased probability of abandoned controls. If sales continue despite unverified or doubtful quality of product, allegations of false claims may follow, leading to fraud and malfeasance. Forced production is rarely discovered by auditors, who are not trained to detect fraud, but it is usually revealed by the testimony of whistle-blowers.

7.1.3.2   Abuse and Threats by Management

Most organizations have a code of business ethics. Although such codes prohibit discrimination based on race, religion, or gender, they often stop there because only these issues are clearly a matter of law. But most such codes do not or cannot eliminate abuse and threats by management. The factory floor bears no more resemblance to the boardroom than does the frontline foxhole to general headquarters. Under production pressures, things can get rough and fur can fly. The issue of fear is one of W. Edwards Deming's (1991a) 14 points for management. Deming understood fear is and continues to be a management tool caused by abuse and threats. Fear engenders an atmosphere of benign neglect at best and at worst, malicious compliance and sabotage. In past generations, it may have been possible to run a productive operation while using sweatshop procedures, but in a modern environment management use of fear on the production line inevitably leads to malfeasance, employee resistance, and whistle-blowing. If higher level management hears of malfeasance via the whistle-blower, it is too late. A formal complaint has been made, the law is notified, and liability looms.

7.2   Management Responsibility

Forensic evidence cited earlier from numerous cases of manufacturing misfeasance shows that almost half of the allegations of misfeasance included operations as a major factor. It also indicates that misfeasance in operations is characterized by poor business practices and systemic process failure.

A quality management system should be so structured as to make it resistant to misfeasance by focusing on business practices and operations management policies. This structure would focus on effective internal controls, business standards of care, liability risk management, employee empowerment, effective management reviews, and closed-loop processes. Let us examine each of these issues in detail.

7.2.1   Effective Internal Controls

It takes little imagination to understand that an ineffective internal control is no control at all. Consider just a few controls of information technology (IT) and of quality for example. Two of the CobIT controls in planning and organization are PO2, *Define information architecture*, and PO5, *Manage the IT investment*. Two of the IT controls in acquisition and implementation are AI2,

Acquire and maintain application software, and AI4, *Develop and maintain procedures* (IT Governance Institute, 2007, pp. 29, 75).

Similarly, Clause 5, *Leadership,* of ISO 9001 requires that senior management communicate to the organization the importance of meeting customer as well as statutory and regulatory requirements. Another requirement of this clause is that top management must provide a framework for establishing and reviewing quality objectives. Still another requires top management to ensure that responsibilities and authorities are defined and communicated within the organization.

Management will have direct impact on the ability of the quality management system to reduce the probability of misfeasance that can lead to systemic process failure. Clause 6, *Planning,* and particularly its requirements to address risk, can be developed into an effective control to reduce the probability of misfeasance. ISO 9004 (ANSI/ISO/ASQ, 2009) offers guidelines for performance improvement and is limited only by the will of the performer to implement its recommendations. However, recalcitrance in establishing effective procedures is not without risk. As pointed out in Chapter 4, in litigation the court may accept ISO 9004 under the umbrella of due diligence and duty of care, in which case the standard is no longer just a guideline but is regarded as an appropriate implementation.

7.2.2 Business Standards of Care

Standards of care refer to the degree of attentiveness, caution, and prudence that a reasonable person in the circumstances would exercise. Failure to meet the standard is negligence. The American Bar Association's (2002) *Model Business Corporation Act,* subscribed to by 24 of the United States, states in paragraph 8.4.2, *Standards of Conduct for Officers,* "an officer shall act in good faith with the care that a person in a like position would reasonably exercise under similar circumstances." In judicial review, the standard of care concept imposes upon management conduct that would be generally considered moral and correct. Conversely, conduct such as employee coercion could be thought beyond a standard of care, thus negligent or indifferent and would invite forensic investigation and legal evaluation for malfeasance.

ISO 9001 can reduce misfeasance because it is an internationally recognized set of good business practices. Some defendant manufacturers have settled out of court rather than go to trial and try to persuade a jury that they are not obliged to use good business practices. In at least one case, the plaintiff used ISO 9001 as its model of good business practices and won the judgment even though the defendant was not certified to ISO 9001 and was not subject to its purview (Broomfield, 1996). This case is discussed in more detail in Chapter 8.

I repeat from Chapter 4 that in the United States, guidance standards such as ISO 9000 and ISO 9004 are viewed as components of a series along with ISO 9001 that can be used to examine issues such as product safety. Guidance documents that are part of a series can be used to establish an organization's due diligence and duty of care and can be used by courts to establish evidence of negligence (Kolka, 2004). As asserted earlier in Chapter 6, the suite of ISO 9000 standards can be considered a set of good business practices. The contribution of the guidelines serves to reduce the occasions of management misfeasance.

7.2.3 Liability Risk Management

Managing risk of liability can be a very big job. One reasoned approach is to view the requirements of ISO 9001 as a set of internal controls and evaluate the risk of each in light of the business and structure of the company. Many safeguards are already in place in most companies, for example, top management review of expenditures and receipt inspection of purchased parts.

Risk management requires accountability. There is an old saying: "Power tends to corrupt and absolute power corrupts absolutely" (Dahlberg, 1887). Power accumulates to those who account to no one. That is why process feedback is important—feedback can be used to create accountability when it is implemented in the organizational structure. Everyone accounts to someone and works to a cost function in proportion to the liability and sensitivity of his or her task (Stimson, 1996).

Internal controls can be an effective element in managing the risk of liability. For example, the following clauses of ISO 9001 can be focused to this effect:

- 5.1, *Leadership and commitment.* Ensure that the management review is staffed and chaired by executive management (plural);
- 8.5.1, *Control of production and service provision.* Require reconciliation of orders and receipts;
- 8.4.2, *Type and extent of control.* Hold the prime contractor explicitly responsible for quality of delivered product;
- 8.3.4, *Design and development controls.* Require validation of delivered product for customer expectations and intended use;
- 8.5.2, *Identification and traceability.* Require traceability of critical parts;
- 7.1.5, *Monitoring and measuring resources.* Ensure that industrial standards are not compromised by the implementation of this control.

The reader is reminded that though these controls are found in every version of ISO 9001, their location within any given version varies. It is incumbent upon the forensic analyst to understand the principles behind each control, and then search for that principle in any governing standard applicable to the litigation.

7.2.4   Employee Empowerment

Employee empowerment has many interpretations, some of them feeble. Effective empowerment was best defined in a now expired Federal QMS:

> Persons performing quality functions shall have sufficient, well defined responsibility, authority, and the organizational freedom to identify and evaluate quality problems and to initiate, recommend, or provide solutions. Mil-Q-9858A (1993)

This kind of empowerment is unfortunately all too rare, but if employees have this authority, then malfeasance is almost impossible. W. Edwards Deming was correct—most employees want to do a good job (Deming, 1991b). Their self-esteem is wrapped up in their trade or craft and they hate to be rushed into slipshod work. They hate to be told to cheat. That is why some people become whistle-blowers. Many people cannot long tolerate actions that they believe are simply wrong and some of them will resist to the point of going beyond their immediate superiors to higher authority—as high as is necessary. You cannot keep an oppressive work environment a secret. Forensic evidence of legitimate employee dissatisfaction is often discovered in emails as well.

Indeed, whistle-blowing is an important writ in law and has its own term, "Qui tam." In such a writ, a private individual who assists a prosecution can receive all or part of any penalty imposed. The phrase comes from the Latin: *Qui tam pro domino rege quam pro se ipso in hac parte sequitur.* The best translation I have found for this phase is "those who pursue this action on behalf of our Lord the King as well as their own" (United States of America District Court, Eastern District of Pennsylvania, 2000).

The subject of oppressive work environments encourages me to admit to an admiration of true whistle-blowers (I exclude malcontents) because their courage is greater than their fear of losing employment. Such tactics as forced production and management abuse and threats are usually hidden to the outside world, but they are very real. Most employees have families to feed and bills to pay. Very often their employer is the only one in town. They are afraid. So when there is wrongdoing in a plant sufficient to outrage employees and to cause whistle-blowing, it divides the company into two: those who side with the whistle-blowers and those who, in fear that the plant will close and move offshore, side with management. Home windows are broken, car tires are flattened, children are frightened, and fights break out. The whistle-blower will live in this fearful environment until the case is resolved, usually years later.

7.2.5   Effective Management Review

A manufacturing system is a complex integration of many activities, each operating at different levels of creation or in different physical locations, but the sum of their efforts being synergistic. In this environment, unwavering management review is necessary for successful outcomes. Quality reviews are one key to synergism because quality is simply good operations. To repeat, Clause 5.1 requires top management to preside over quality reviews in order to ensure sufficient authority to make necessary decisions. But in some companies only the quality assurance officer presides over quality reviews, which may meet the requirement but does not meet the intent.

The absence of top management can lead to an ineffective review and provide an opening for misfeasance. This risk is reduced by having several executive managers on boards of review. A strategy of management teams in a collective review process ensures the necessary authority and increases the breadth of managerial expertise. It also provides the transparency needed for self-monitoring among peers.

While auditing the Metro Machine ship repair yard of Chesapeake, VA, some years ago, I observed that weekly progress meetings were conducted for each ship in repair. All the meetings were chaired by the president of the company, with the vice president for production and all foremen and key personnel in attendance (Stimson, 1993). The progress meetings conducted in this shipyard remain in my mind as a model of effective performance management. All the necessary dynamics were present: the strategic force, the tactical force, and the fundamental controls of responsibility, authority, and accountability.

From a forensic perspective, evidence of the presence of several executive managers in management reviews demonstrates their effectiveness, the strengthening operations, and an increasing the probability of compliance to contractual obligations.

7.2.6 Closed-Loop Processes

ISO 9000 (ANSI/ISO/ASQ, 2005) defines an *organization* as a group of people and facilities with an arrangement of responsibilities, authorities, and relationships. Responsibilities and authorities are internal controls. Hence, an organization is a set of internal controls with a defined and synergistic relationship and control systems theory is appropriate to it. Every control system must have a feedback structure, which is necessary for stability but does much more than that. It permits comparing what is to what should be and can be used for verification of every kind of activity within the process. There is feedback in a correctly structured organization. For example, line managers feed back their progress to top management in strategic terms in order that they are understood. The classic structure of the closed-loop feedback system is shown in Figure 7.1.

The output of the controller is the control law—a rule or set of rules used to ensure that a control works according to plan. For example, consider ISO 9001, Clause 5.1.2, *Customer focus*: "Top management shall ensure that customer requirements...are determined, understood, and consistently met." Relative to

Figure 7.1 A classic closed-loop system.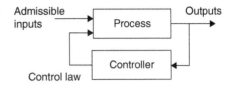

Figure 7.1, Clause 5.1.2 is a control and its requirements are the control law for that process.

Envisioning organizational processes as closed-loop control systems aids forensic review to verify the implementation of internal controls. Organizational feedback is identified with simple questions. Who is reporting what, to whom, and when? Are action items assigned, conducted, and verified? Who is responsible? Who is accountable? Conversely, a process with no feedback is called an *open-loop process*. There is little effective monitoring in an open-loop process, which opens the door to negligence, misfeasance, and process liability (Stimson, 2001).

In many contracts, an effective performance management standard is required. One such standard is ISO 9001, which is both contractual and universal. If you win a contract based on certification to this standard, you must by law use it effectively. Millions of dollars in liability and damages are at risk as a result of ineffective management. Because it creates a closed-loop management system, ISO 9001 enables effective operations.

References

American Bar Association (2002). *ABA Model Business Corporation Act*. http://www.americanbar.org/content/dam/aba/administrative/business_law/corplaws/model-bus-corp-laws-w-o-comments-2010.authcheckdam.doc. Accessed March 28, 2017.

ANSI/ISO/ASQ (2005). *ANSI/ISO/ASQ Q9000-2005: Quality Management Systems—Fundamentals and Vocabulary*. Milwaukee, WI: American National Standards Institute and the American Society for Quality.

ANSI/ISO/ASQ (2009). *ANSI/ISO/ASQ Q9004-2009: American National Standard: Managing for Sustained Success*, Table A3. Milwaukee, WI: American National Standards Institute and the American Society for Quality.

ANSI/ISO/ASQ (2015). *ANSI/ISO/ASQ Q9001-2015, Quality Management Systems—Requirements*. Milwaukee, WI: American National Standards Institute and the American Society for Quality.

Arora, J. S. (1989). *Introduction to Optimum Design*. New York: McGraw Hill, pp. 4–11.

Boehm, B. W. (1984). "Verifying and Validating Software Requirements and Design Specifications." *IEEE Transactions, Software Engineering*, January, pp. 75–80.

Broomfield, J. R. (1996). "Lawyers Wise Up to ISO 9000." *Compliance Engineering*, January, pp. 32–39.

Dahlberg, J. E., Lord Acton (1887). The phrase is in a letter from Lord Acton to Bishop Mandrell Creighton. *The Oxford Dictionary of Quotations*, Oxford University Press [1992], p. 1.

Deming, W. E. (1991a). *Out of the Crisis*. Cambridge, MA: Massachusetts Institute of Technology, Center for Advanced Engineering Study, p. 59.

Deming (1991b), p. 83.

Federal Acquisition Regulations (1995). *FAR 46-405(a), Subcontracts*. http://farsite.hill.af.mil/reghtml/regs/far2afmcfars/fardfars/far/46.htm#P185_28365. Accessed September 1, 2017.

Federal Acquisition Regulations (1996). *FAR 52-246-2(b), Inspection of Supplies*. http://farsite.hill.af.mil/reghtml/regs/far2afmcfars/fardfars/far/52_246.htm#P8_708. Accessed September 1, 2017.

Goldratt, E. and Cox, J. (1986). *The Goal: A Process of Ongoing Improvement*. New York: North River Press, p. 76.

Imai, M. (1997). *Gemba Kaizen*. New York: McGraw Hill, p. 77.

IT Governance Institute (2007). *Control Objectives for Information and Related Technology 4.1*. Rolling Meadows, IL: Governance Institute, pp. 29–168.

Kolka, J. W. (2004). *ISO 9000: A Legal Perspective*. Milwaukee, WI: ASQ Quality Press, p. 61.

MIL-Q-9858A (1993). *Military Specification: Quality Program Requirements*. Washington, DC: United States Department of Defense, p. 2.

Stimson, W. A. (1993). *Metro Machine Corporation: A Malcolm Baldrige Quality Award Assessment*. School of Engineering and Applied Sciences, University of Virginia (28 June 1993). Report to the Systems Engineering Department.

Stimson, W. A. (1996). *The Robust Organization: Transforming Your Company Using Adaptive Design*. Chicago, IL: Irwin, p. 60.

Stimson, W. A. (2001). *Internal Quality Auditing*. Chico, CA: Paton Press, pp. 247–248.

United States Code 31 (2007). §3729 et seq., *False Claims Act*, January 3, 2001. https://www.law.cornell.edu/uscode/text/31/3729. Accessed September 19, 2017.

United States Congress (2002). *H. R. 3763: The Sarbanes–Oxley Act of 2002*. Title IV, Section 404: *Management Assessment of Internal Controls*. Washington, DC: United States Congress. https://www.congress.gov/bill/107th-congress/house-bill/3763. Accessed September 19, 2017.

United States Department of Justice (2008). *Fact Sheet: President's Corporate Fraud Task Force Marks Five Years of Ensuring Corporate Integrity*. https://www.justice.gov/archive/opa/pr/2007/July/07_odag_507.html. Accessed September 15, 2017. In 2009 the "Corporate Fraud Task Force" title was changed to the "Financial Fraud Task Force."

United States Department of Justice Civil Division (2010). Civil Division press releases, https://www.justice.gov/news. Accessed September 19, 2017.

United States of America District Court, Eastern District of Pennsylvania (2000). Civil Action Number 94-7316, August 2000. https://www.paed.uscourts.gov/documents/opinions/00D0664P.pdf. Accessed September 1, 2017.

U.S. Code of Federal Acquisition Regulations (FAR) (2014). Title 48, 52.222-36, *Affirmative Action for Workers with Disabilities.* http://farsite.hill.af.mil/reghtml/regs/far2afmcfars/fardfars/far/52_220.htm#P625_106241. Accessed September 15, 2017.

8

Legal Trends to Process Liability

Historically and currently, customer injury or harm that occurs because of a defective product or service is viewed within a context of product liability. A person is harmed by a product or service and the existence of a defective unit is necessary to that complaint. This position is held even if there are hundreds of injuries. Then hundreds of bad units must be found. The present approach to large-scale product nonconformity is in the form of class action suits. Individuals, perhaps scattered all over the country, come forward one by one and present their individual evidence.

8.1   An Idea Whose Time Has Come

The idea of process liability is, at present, just that—an idea. The possibility that a system may be producing hundreds, if not thousands, of nonconforming units unknowingly because no one is looking is still beyond consideration.

Although the idea of process liability is not yet explicit in the judicial system, courts are moving in that direction in that they are increasingly concerned with management use of good business practices (Boehm & Ulmer, 2009a). A cause and effect relation between process and product is not yet established in law, but we are getting there. Decisions in class action suits indicate that the law now views manufacturing and service as a set of interrelated processes. Let us examine a few cases in which the trend to process liability is perceptible in court decisions.

Forensic Systems Engineering: Evaluating Operations by Discovery,
First Edition. William A. Stimson.
© 2018 John Wiley & Sons, Inc. Published 2018 by John Wiley & Sons, Inc.

In industrial engineering, causes of variability in product or service characteristics fall in two categories: common and special (Deming, 1991a). Common causes are random and inherent to the system because no system or design is perfect. Causes that are external to the design of the system and cause the system to be unstable are called "special." Such causes can and must be identified and eliminated in order to reestablish the stability and statistical properties of the system.

Sometimes common causes can increase to the point where they may destabilize the system, perhaps over an extended period (Deming, 1991b). Whatever the name, causes that create unstable operations are the responsibility of management, so that in the event of systemic failure, which can usually be identified from evidence, it is but a short step for an attorney to prove misfeasance and perhaps even malfeasance and false claims. Systemic failure suggests misfeasance and a reasonable conclusion of process liability.

8.2 Some Court Actions Thus Far

Some defendant manufacturers have settled out of court rather than go to trial and try to persuade a jury that they are not obliged to use good business practices. In a later section of this chapter, we will discuss the case in which the plaintiff used ISO 9001 as its model of good business practices and won the judgment even though the defendant was not certified to ISO 9001 and was not subject to its purview.

I repeat from Chapter 4 that in the United States, guidance standards such as ISO 9000 and ISO 9004, although noncontractual, are part of a series that with ISO 9001 that can be used to examine issues such as product safety, due diligence, and duty of care (Kolka, 2004a). Thus, a plaintiff could argue that the suite of ISO 9000 standards can be considered a set of good business practices and one can reasonably expect that a company would adopt similar practices as part of its corporate responsibility and to reduce the possibility of management misfeasance.

Traditionally, we assume that if a product is offered for sale, it should be fit for intended use (Boehm & Ulmer, 2009b). This is an important assumption in ISO 9001, which has several requirements for the intended use of product. Beyond this, the term itself, *intended use*, has an important implication with respect to product reliability. Chapter 10 expands on this relationship because of its significance in forensic analysis.

Courts now look beyond the defect of a product to a company's quality assurance procedures to determine their ability to prevent and control risks. This may include an assessment of the company's effectiveness in all processes that may affect the quality of the unit: design, marketing, manufacturing, shipping, service, and supplier selection. Existing quality management systems (QMSs) provide a formal mechanism for most of the facets of due diligence demanded by the courts.

Exposure to liability exists throughout the product life cycle from conception through design, production, warranty, time in-service, to end of life. Product designers must consider the ultimate use of the product for safe use. In production, the product must meet specifications that include safety features against hazards that cannot be designed out—lawn mowers, for example. In marketing, warranty and advertising claims must accurately reflect product life cycle, safety, and performance expectations. These traditional areas of liability closely parallel the requirements of ISO 9001, which are excellent descriptions of good business practices and as such offer guidance that could be useful in liability litigation.

8.2.1 QMS Certified Organizations

In a 1999 lawsuit involving ISO 9001, *Case International Harvester Cotton Picker Fire Products Liability Litigation,* precedents were established that recognize ISO 9000 as a source of good business practices, incumbent upon performers (Kolka, 2009a).

Plaintiff's attorneys requested that the defendant turn over ISO 9001 documents relating to fire problems with the Case IH cotton picker, arguing that "Defendant's ISO 9001 quality assurance manual and all other ISO 9001 documents relating to these fires are discoverable." On June 21, 1999, the US District Court for the Eastern District of Arkansas ordered the defendant to turn over materials from its ISO manual to plaintiff's attorneys for evaluation.

Plaintiffs also argued: "If defendant failed to comply with its own ISO 9001 certification, that information is relevant to negligence and punitive damages. If defendant complied with its own ISO 9000 procedures, but these procedures are themselves inadequate, plaintiff is entitled to discover that information because it also would be relevant to negligence."

Specifically, plaintiff wanted to examine corrective action and management review minutes to determine if Case had followed its own ISO 9001 corrective action procedures and if not, why not? But as a corporate attorney had been present during an in-house correction active review, Case argued that the review minutes fell under attorney–client privilege and therefore did not have to be turned over in legal discovery. Nevertheless, the US District Court Judge ordered the defendant to turn over the minutes to plaintiff. Subsequently, the defendant petitioned the US Eighth Circuit Court of Appeals to overturn the judge's order.

On January 24, 2000, Judge Thomas Eisele of the Appeals Court denied defendant's petition and upheld the district court judge's ruling to turn over the minutes of the corrective action review. Having lost its appeal at the Circuit Court level, in April 2000, Case filed a petition for a Writ of Certiorari with the US Supreme Court, asking the higher court to review the decision of the lower courts.

On June 5, 2000, the US Supreme Court denied the Writ, thereby upholding the Eighth Circuit Court decision that Case had to turn over its minutes of management and corrective action reviews to plaintiff's counsel. Thus, at all judicial levels, the conclusion is that an attorney–client privilege does not apply to ISO 9001 records and documentation just because an attorney was present during the corrective action review.

It is useful to review some precedents established by the Case IH litigation, as perceived by attorney James Kolka.

1) A company can be liable if it does not follow its own QMS and this failure results in property damage or injury.
2) A company can be liable if it does follow its own QMS, but its procedures are inadequate, resulting in property damage or injury to a customer or user.
3) Representing universally recognized good business practices, ISO 9000 documentation, records, and minutes can provide valuable information and ideas on litigation strategy in the event of a law suit.
4) ISO 9000 documentation may not be shielded from legal discovery by claiming attorney–client privilege.
5) Information obtained from customer complaints, nonconformities, and corrective action generally should result in design review and possibly a production process review. All relevant elements of an ISO 9001 QMS should be examined when addressing a customer complaint.

Judge Eisele ruled that, pursuant to Federal Rules of Civil Procedure 401 and 402, "all 'relevant' evidence is admissible (Kolka, 2009b). Evidence is 'relevant' if it has a tendency to make the existence of any fact that aids in determining an action more or less probable than it would be without the evidence."

He further noted that "the Eighth Circuit has consistently held that compliance or noncompliance with American National Standards Institute (ANSI) standards is admissible in strict liability and/or negligence cases to prove a design defect." It should be noted that ISO 9001 is an ANSI standard.

These rulings support the notion that private industry has a legal obligation to use good business practices in operations that provide goods and services to the public.

Judge Eisele ruled also that the presence of an attorney in a corrective action or management review does not constitute attorney–client privilege. This ruling does not directly bear upon good business practices, but it does reduce the opportunity to shield poor practices from public scrutiny.

8.2.2 QMS Noncertified Organizations

In a 1996 issue of *Compliance Engineering*, author John Broomfield (1996) discusses an unnamed case in which the collapse of a chaise longue caused the plaintiff to reinjure a knee that was healing from reconstructive surgery. The injured plaintiff blamed the manufacturer, asserting a leg of the chair was improperly fastened and that assembly instructions were incomplete and confusing. Plaintiff's counsel argued that the manufacturer had ignored the ANSI/ASQ 9001 standard and did not document its own training or inspection program.

Interestingly, the manufacturer was not registered to an ISO standard, yet it was challenged for ignoring quality standards. In deposition, the defendants were questioned by plaintiff's attorney using clauses from ISO 9001 that addressed design control, training, and inspection and testing. The case was settled out of court with the plaintiff receiving $55,000 from the manufacturer.

Since the case did not go to trial, we cannot know whether the judge would have permitted plaintiff's attorney to pursue this line of questioning. But clearly the defendants did not want to risk the possibility that questions might be allowed using clauses from ISO 9001. They chose to settle out of court rather than risk possible embarrassment in a trial where their business processes would fail when compared to processes from an international quality standard.

Operations can go astray in many ways but often they will occur in three general areas: verification and validation, resource management, or process management. Processes within these areas are usually required by law or standard to have well-defined procedures. If a process departs from accepted procedure, or if a procedure itself is absent or remiss, the result is a nonconforming system and if continued, systemic product failure follows. The processes listed here are critical and representative. If nonconformity is sustained, management misfeasance is indicated. Increasingly in time, court decisions reflect the consideration that management has a responsibility to ensure not only that the products and services are as advertised but that the processes used to provide them are properly operated.

References

Boehm, T. C. and Ulmer, J. M. (2009a). Product Liability Beyond Loss Control—An Argument for Quality Assurance. *Quality Management Journal*, vol. 15, p. 11.

Boehm and Ulmer (2009b), p. 10.

Broomfield, J. R. (1996). Lawyers Wise up to ISO 9000. *Compliance Engineering*, January, pp. 32–39.

Deming, W. E. (1991a). *Out of the Crisis*. Cambridge, MA: Massachusetts Institute of Technology, p. 314.

Deming (1991b), p. 411.

Kolka, J. W. (2004a). *ISO 9000: A Legal Perspective*. Milwaukee, WI: ASQ Quality Press. Attorney Kolka, PhD, JD, is a legal expert in compliance regarding ISO 9001, ISO 14001, CE Marking, product liability, and other management systems. p. 61.

Kolka, J. W. (2009a). *ISO 9001 Case IH Cotton Picker Fire Products Liability Litigation #2*. http://jameskolka.typepad.com/international_regulatory_/2009/03/iso-9001-case-ih-cotton-picker-fire-products-liability-litigation-2.html. Accessed September 20, 2017.

Kolka, J. W. (2009b). *ISO 9001 Lawsuits & Document Control #2*. http://jameskolka.typepad.com/international_regulatory_/2009/03/index.html. Accessed September 20, 2017.

9

Process Stability and Capability

I asserted in Chapter 8 that common causes can increase to the point where they may destabilize the system, be it manufacturing or service. Causes that create unstable operations are the responsibility of management and in the event of systemic failure, misfeasance and process liability may follow. Hence, process stability is essential to the provision of products and services that conform to customer requirements. This assertion shall be proven in later chapters, but for now, we must understand what stability means.

9.1 Process Stability

In the Preface, I quoted Kalman's definition of a dynamical system as an abstract mathematical concept defined by a set of vector spaces: admissible input functions, single-valued output functions, system states, and a state transition function (Kalman et al., 1969). In principle the definition applies to stochastic

Forensic Systems Engineering: Evaluating Operations by Discovery,
First Edition. William A. Stimson.

as well as deterministic systems. In this context, we begin with the classical definition from Liapunov's general theory on stability, taken from Siljak (1969):

An equilibrium state x_e of a free dynamic system is stable if for every real number $\varepsilon > 0$ there exists a real number $\delta(\varepsilon, t_0) > 0$ such that

$$|x_0 - x_e| \le \delta \ implies$$

$$|x(t, x_0, t_0) - x_e| \le \varepsilon, \quad for \ all \ t \ge t_0$$

If x_e is not stable, then it is said to be unstable.

In plain English, if a δ neighborhood about the equilibrium state x_e can be found for the initial state, x_0, such that the state trajectory is bounded within an ε neighborhood of x_e for all time, then x_e is stable. The geometry of this condition and the random trajectory of x_t is shown in Figure 9.1, on the left.

9.1.1 Stability and Stationarity

With respect to stochastic systems, an analogous geometry of the covariance, γ_k, of an autoregressive systems, $x(t)$, is shown in Figure 9.1, on the right. The rationale derives from the definition of weak stationarity. Ross (1989) defines weak stationarity as follows: a stochastic process $x(t)$ is stationary in the weak sense if its expected value is constant and its autocorrelation depends only upon the interval of the correlation, τ.

$$E\{x(t)\} = \mu \qquad E\{x(t + \tau)x(t)\} = R(\tau) \tag{9.1}$$

Because μ is constant, so also is $\sigma^2(t)$ constant. Thus, the boundedness of γ_0. If the stochastic process is discrete, then the index is not τ but k, an arbitrary

(a)   (b)

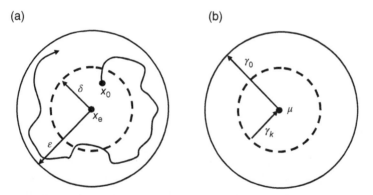

Figure 9.1 Phase plane geometry of dynamic system stability. (a) A bounded trajectory in state space near a stable equilibrium point, X_e and (b) Covariance diminishing with increasing k, of a distribution in equilibrium with weak stationarity.

epochal index used in Figure 9.1. As k increases, γ_k decreases if the polynomial, $\gamma(B)$, converges as indicated by the *autocovariance generating function (agf)* of Box and Jenkins (1976a):

$$\gamma(B) = \sum_{k=-\infty}^{\infty} \gamma_k B^k \qquad (9.2)$$

where B is the backward shift operator. The circular bounds on γ_k decrease in radius to degeneracy as k increases. The operator B is defined by $Bz_t = z_{t-1}$. In time series analysis, B is treated as an algebraic quantity and dummy variable, as is D in differential calculus. Hence, one can solve for the roots of an nth order polynomial in B.

A discrete, stochastic process is stable if its eigenvalues lie on or within the unit circle (Masanao, 1987). Eigenvalues represent the magnitude of the variance of a set of data and can be found from the correlation matrix of input observations taken from a stochastic process. However, there is another, less strenuous approach in my view, using estimates of process parameters and correlations. The estimates reveal stationarity, and stationary processes are stable.

Lütkepohl (2005) says that stability implies stationarity, so that the stability condition (location of the eigenvalues) is often referred to as the *stationary condition* in the time series literature (italics of the author). The concepts of stability and of stationarity are so entwined in stochastic processes that the two terms are often used interchangeably. However, they are defined differently, and so they are not the same thing. Stability is a property of motion, whereas stationarity is a property of randomness. The objective of an internal control is to maintain the stability of a process. Therefore, stability in the classical sense must be defined in terms that can be established by data. The following definition achieves this purpose and is shown to conform to accepted systems and statistical theory.

Definition
A serially dependent process is stable if its autocovariance generating function, $\gamma(B)$, converges (Stimson, 1994).

9.1.2 Stability Conditions

The definition of stability says nothing about how *fast* $\gamma(B)$ converges. In stability theory, the process is allowed to return to an equilibrium point in infinite time (DeRusso et al., 1966), but statisticians recognize that a very slowly decaying $\gamma(B)$ indicates nonstationarity (Bowerman & O'Connell, 1979). How is this contradiction between control and statistical theory reconciled?

The answer lies in the notions of *homogeneous nonstationarity* and *marginal stability*. Abraham and Ledolter (1983) say that homogeneous nonstationarity infers a changing mean describable by low-order polynomials with nonconstant coefficients. Box and Jenkins (1976b) say that this property indicates roots on the unit circle. However, roots on the unit circle do not imply instability in control theory, but they rather imply marginal stability. Marginal does not mean temporary. A sinusoidal process is marginally stable forever, as long as the roots stay on the unit circle. Thus, the compromise. One must distinguish between stability and nonstationarity in the case of unit roots. In particular, as stated by Lütkepohl, instability does not imply nonstationarity. The definition of stability posed here is satisfying for another reason. Stationary processes have constant distributions. Gilchrist (1976) calls this "statistical equilibrium," pointing out the equilibrium state enables determination of control limits in quality control because such limits require a constant process variance. I shall show later that the phrase "in control" as used in Shewhart control charting, is formally tied to process stability.

In summary, a stochastic, observable system with bounded output is stable. It may be nonstationary, which means that neither its first or second statistics nor its capability can be determined. If the system output is being charted and trends or other nonrandom behavior is evident, their causes must be detected and corrected before the control limits are exceeded. If the system output is not being charted, then the volume of the output that is nonconforming is certain but unknown.

9.1.3   Stable Processes

Simply put, engineering is the practical application of physical and mathematical theories. So what is the practical application of stability theory to modern industrial processes? Walter A. Shewhart answered this question in a seminal volume descriptively entitled *Economic Control of Quality of Manufactured Product* (1931). In doing so, he established some basic assumptions about the statistics of manufacturing that address the *stability* of the productive process. Shewhart recognized that the industrial process is stochastic and the critical value of a product's characteristics varies randomly with time (Shewhart, 1931a). The critical value defines the quality of the product and was often called the critical quality characteristic, but the newer term is "key characteristic" (Aerospace Standard 9100, 2004), or "key value" for brevity. By stability Shewhart meant that this quality variation, though random, has a distribution with constant mean and variance, and bounded magnitude.

(There is some debate on the criticality of "key characteristic," with some sources defining a key characteristic as that variation within tolerance that could affect the effectiveness of the product, but which is not critical. They prefer the term "critical characteristic," defined as that variation within tolerance that could affect user safety or be noncompliant to regulations. Unit criticality may be an important issue in litigation, but for brevity in this book, I use the term "key value.")

Left to itself, the process variation may not be stable and Shewhart devised a means to attain, measure, and maintain quality stability called "statistical process control (SPC)." Classical SPC focuses on determining whether the distribution of product randomness is stable. In its simplest form, suppose that x_t is the random measurement of the key value of a manufactured product. Assume further that x_t is normally distributed with mean μ and known variance σ^2. If the production process is stable, then μ will be close to a specified target value μ_0. According to Shewhart, $\bar{x}$ (x-bar), the mean of k items sampled at any instant of production, should lie between control limits

$$CL = \mu_0 + \frac{3\sigma}{\sqrt{k}} \tag{9.3}$$

with a 0.9973 probability if the process is stable. Continual sampling of $\bar{x}$ verifies the stability of the process and remedial action is taken if the control limits are violated. A measured standard deviation is denoted by the letter "s".

Following the design of the unit, a process, too, must be designed that can produce the unit with its design value, which becomes the key value. There are two considerations with respect to this process: first its stability; second its capability. Process stability refers to the ability of the process to produce the same unit over and over again, all with their key value at the design value or within acceptable limits. As no process is perfect, the key value will vary randomly, but if that variation is bounded within acceptable limits, then the process is said to be stable.

Figure 9.2 shows how a random variable behaves when distributed normally. Although the natural variation of a given key value may not be normally distributed, there is no loss in generality in assuming so and indeed, the way stability is measured ensures that the key value *is* distributed normally.

The horizontal axis is measured in multiples of the process standard deviation, which by convention is called σ (sigma). The area under the curve

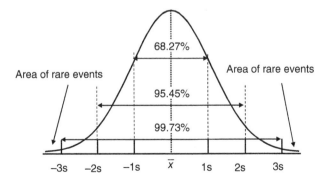

Figure 9.2 Normal distribution of a random variable.

indicates the probability that a given event will be within that spectrum. Figure 9.2 shows that in a normal distribution, 99.73% of the measured key values, $\bar{x}$, will fall within ±3 standard deviations of the process mean value. Hence, only 0.27% of the values exceed this range, and because the normal distribution is symmetrical, the probability of being either less than -3σ or greater than $+3\sigma$ is 0.00135, a very small probability and hence a "rare event."

9.1.4 Measuring Process Stability

Process stability is measured by taking random samples of the units being provided, measuring each key value, and then plotting the values as a time series. However, individual key values are not plotted. Instead, the sampling is done in small subgroups of 2–5 units each and the average value, $\bar{x}$, of the subgroup is plotted. The reason for taking small subgroups and averaging their key value is that averages are approximately normally distributed (Grant & Leavenworth, 1988a). In this way, the normal distribution can be used to define the areas of stability and rare event for any process.

A typical plot of average key values as a time series is shown in Figure 9.3. The upper and lower control limits (UCL, LCL) indicate the boundaries of 3 standard deviations. The grand average value, $\bar{\bar{x}}$ (average of the averages), is also plotted in order to provide a reference to indicate any trends. As long as the variation of the key value stays within the control limits, no action is taken. This variation constitutes the inevitable band of chance causes described by Shewhart. The variation is bounded and indicates that the process is stable. The process is said to be in statistical control (Shewhart, 1931b), or in common usage, simply "in control."

There is nothing sacrosanct about 3 standard deviations. Other industries might choose other probability limits. For example, a hospital intensive care unit that is monitoring the heart rate of its patients might choose 2 standard deviations as stability limits, preferring the expense of chasing false alarms to the possibility of ignoring a patient in extremis.

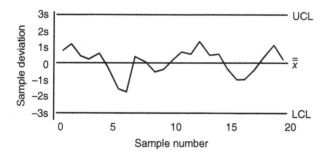

Figure 9.3 A control chart of a characteristic (key) value.

Nevertheless, the industrial convention is to use ±3 standard deviations as control limits. Lower limits would lead to the expense of pursuing false alarms—those external system disturbances that seem to exist but do not, being simply part of the natural system variation of common causes. Higher limits would lead to ignoring very real system disturbances created by exogenous forces.

9.2   Process Capability

To say that a process is stable simply means that its variation is bounded. This variation has a constant mean value and a standard deviation that can be measured. Yet, stability says nothing about how good the process is. The process mean value may be nowhere near the design value. Process stability verifies consistency, but says nothing about whether the process is able to meet customer requirements. A measure of performance is needed that can determine just how good the process is—its capability, if you will.

The Dictionary Britannica defines *capability* as the state or quality of being capable; or susceptible to some particular use; or a condition that may be developed or improved. Relative to systems, all three definitions pertain. In systems engineering, capability is defined as the ratio of acceptable deviation to process deviation. This definition is at the same time a doubly informative measure of performance because once you know the capability, you can identify the effect of improvement strategies.

9.2.1   Measuring Capability

Process capability is determined separately and after stability is established. Stability must be achieved first because the process statistics must be constant and identified in order to determine process capability.

Knowing that no operational process is perfect and some variation will always exist, design engineers will consider how much variation about a design value a process may assume and still meet customer requirements. After all, how precise is precise? If you want to drill a hole to a depth of ¼ inch, is the design value 0.25? 0.250? 0.2500? Can the hole have a depth of, say, 0.245–0.255 inches and still meet requirements? The acceptable range about the design value is called the specification limits, or "specs," and defines the span of off-design key values that can still meet customer requirements.

Therefore, in the design of a unit and the design of a process that can produce the unit, we have two ranges of variation to consider. The first is the acceptable variation of the key value of the unit itself, integrated into the unit design as specifications. This variation becomes acceptable when all values within this range will result in a product or service that meets customer requirements.

The second is the actual variation in the key value that will occur because the process that makes the unit is not perfect. This variation must be bounded within ±3 standard deviations of the process mean value in order to establish process stability. Ideally, the process mean value will be at or near the design value.

Process capability is then measured as the ratio of acceptable variation to process variation. Acceptable variation lies between the specification limits. Process variation is the variation inherent in the operational process. If the process is designed well, the process mean value will be at or near the design value and the process variation limits will be less than the specification limits. The greater the ratio, the higher the process capability and the greater the volume of acceptable output.

Process variability is always expressed in standard deviations, σ, in which ±3σ are the control limits. If the specification limits also happen to be ±3σ, then you get a ratio of unity (one). This ratio is called the *process capability*, or C_{pk}. It means that over 99% of the process output is within spec. There is less than 0.3% waste. Though this sounds impressive, it really is not. Japan attained an industrial average C_{pk} of 1.33 nearly 30 years ago (Bhote, 1991). The distribution on the left in Figure 9.4 is a graphic display of this process capability.

The distribution on the right in Figure 9.4 shows the event in which the process has too great a variability—too much of its output exceeds specification. The specification width is about ±2σ, so the C_{pk} is only 0.67. This means that about 5% of the product exceeds specification and represents waste.

If the process mean value is not at or near the design value, then process capability is measured from the process mean value to the nearest control limit. Obviously, there will be some loss in capability, but the result may still be acceptable, depending on the desired capability.

Whether a capability index is "good" or not is rather arbitrary and various industries often establish an expected capability for its members. For example, Eaton Corporation retains a supplier production part approval process

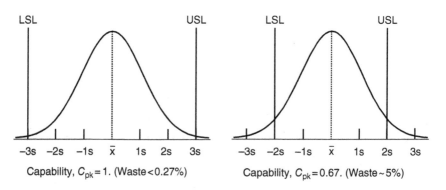

| LSL | | | USL | | LSL | | | USL |

−3s −2s −1s $\bar{x}$ 1s 2s 3s   −3s −2s −1s $\bar{x}$ 1s 2s 3s

Capability, C_{pk} = 1. (Waste < 0.27%)   Capability, C_{pk} = 0.67. (Waste ~ 5%)

Figure 9.4 Process variation superimposed on specification limits.

agreement (Advanced Product Quality Planning (APQP) and Production Part Approval Process (PPAP), 2015) that requires a $C_{pk} \geq 1.67$ for critical parts and $C_{pk} \geq 1.33$ for noncritical parts. To put these indices in context, the goal of a Six Sigma program is the equivalent of a $C_{pk} = 2$, or fewer than 4 defective parts per million (Harry, 1997).

9.2.2   A Limit of Process Capability

In his paper, Miller discusses a corollary issue of process liability—in the interest of justice, there must be a limit to it. He writes: "In process liability there should also be a limit to the liability and this limit should be based on the concept of fault. If the defendant has shown that he has utilized the best available methods, he should not be liable in process liability" (Miller, 1970).

Miller's justification for placing a limit on process liability where "best available methods" are used is in close accord with the engineering sense of a statistical rare event in operations. Even a good system can produce a defective unit, although with very low probability. Earlier, in Section 9.1, a process was defined as stable if the key values of its units are within control limits. Process capability can be measured only when the process is stable. If the process becomes unstable, its variance cannot be measured, and hence neither can its capability be measured.

9.3   The Rare Event

9.3.1   Instability and the Rare Event

An industrial operation is stochastic—its quality varies randomly in time. Therefore, the forensic analyst uses statistical methods to delineate that nonconformance which is random from that which is systemic. A stable system may suffer random nonconformity. An unstable system suffers systemic nonconformity. There are clues to whether or not a pattern of deviation is occurring in the process output and, in particular, the analyst must be aware of a "rare event" deviation because it limits the utility of the evidence. The forensic analyst must be able to identify systemic failure because random failures, no matter how many, can be argued to be rare events, given an infinite production, thus forcing a jury to decide on a highly technical matter that it may well be untrained to understand.

To be fair, it must be recognized that there is no such thing as a perfect system of production or service. Walter Shewhart, the "father of modern quality control" (American Society for Quality, 2016), developed a definitive argument to this effect in his seminal book, *Economic Control of Quality of Manufactured Product*, which is summarized by Grant and Leavenworth (1988b) as follows: "Measured quality of manufactured product is always subject to a certain amount of variation as a result of chance. Some stable system of chance causes is inherent in any particular scheme of production and inspection. Variation within this stable pattern is inevitable. The reasons for variation outside this stable pattern may be discovered and corrected."

Therefore, it is possible that injury or damage can result from a faulty product or service provided by excellent systems of operations and duty of care. Forensic analyses are required in making decisions about the performance of such systems. Forensic systems engineering must accommodate both the legal precedents in governance of operations and the technical advances in systems engineering.

Although no system of production or service is infallible, it is nevertheless possible to maintain a very high order of effectiveness through good business practices and due diligence. In short, it is possible to attain a quality management system in which failure is a rare event. In such an excellent system, if a mishap occurs then injury compensation may be reasonable, but punitive damages based on fault would not be appropriate because there is no fault. The bases of this conclusion are two: the legal idea of process liability, as argued by attorney Miller, and the engineering idea of a rare event nonconformity. We should note at this point that while Shewhart was speaking of manufacturing, it is recognized today that his argument applies to processes in general and is equally valid when applied to service industries (ISO 9001, ASQ/ANSI/ISO, 2015).

9.3.2 Identifying the Rare Event

The rare event philosophy in quality engineering recognizes that no system of operations is perfect and that there will always be nonconformity which, however, will be rare when best practices are vigilantly maintained. I again repeat the earlier quote: "Some stable system of chance causes is inherent in any particular scheme of production and inspection." By industrial convention, the stable system of chance causes in a process corresponds to that variation bounded within ±3 standard deviations of the process mean value. Variation beyond this range, which has a probability of less than 0.0027, is assumed to be caused by an external disturbance that must be identified and corrected. Therefore, in the context of operations, the occurrence of a defect is a rare event if the probability of it happening is less than 0.0027, or 0.27%, of a given operational volume. The areas of rare event in the normal distribution occur in the tails, as indicated in Figure 9.2.

There is no mathematical law here; the statistic comes from an economic convention that must be made because there is no such thing as a perfect system of operations. If an operation yields a nonconformity, a judgment must be made on the cause and whether it can be economically pursued.

A process in which the variation is bounded within ±3 standard deviations of its mean value is considered in control. Nevertheless, any system is subject to natural variability according to Shewhart (1931c). Since no system of operations is perfect, we must recognize that humans and their machines are not infallible and nonconformity will happen. Therefore, it seems reasonable to conclude that if a person is injured by the failure of a unit and evidence points to a system in control with a rarity of mishap, then a "rare event" plea is an appropriate defense. To repeat, compensation for injury may be reasonable, but damages derived from fault are not appropriate from an engineering perspective.

Courts have tended to agree, recognizing that many production and service processes are complex and can involve thousands of tasks. No quality control is capable of catching all defects. Whether the performer will be liable in negligence to persons injured will depend upon a determination of whether the process and quality control procedures were conducted with "reasonable care" (Owen et al., 2007a).

The question in law is whether a unit is "defective." The general legal view is that perfect safety is not possible and can cost too much in dollars and in usefulness. Courts routinely assert that manufacturers are not insurers of product safety and do not have a duty to sell only the safest possible products. Since absolute safety is not the rule, the question is one of balance between safety, usefulness, feasibility, and cost. In any given litigation, the court must decide, "How safe is safe enough?" (Owen et al., 2007b).

Within the context of systems engineering, the rare event plea would be valid only if the operational system is in control. Among other things, "in control" requires that correct procedures are defined and disseminated, in use at the time of the mishap, and operations personnel are adequately trained.

9.4   Attribute Testing

Some key characteristics have no dimension or, having dimension, are more efficiently examined for stability by measuring them as attributes because it can be less expensive to measure an attribute than to measure a dimension. The simplest attribute is a binary characteristic—either the unit conforms or does not conform to specifications. A typical criterion is pass/fail or go-no go. For example, given a coffee service, a cup of coffee is either served properly or it is not. Binary attributes can be modeled by a binomial distribution that, given adequate sample size, approximates a normal distribution. All the arguments presented in this chapter apply to key values as attributes. It is shown in Chapter 14 that very often a problem can be framed so that the variation of key values, although in dimensional form, can be measured as attributes, thereby simplifying evidence that may be presented to a jury.

In attribute testing, processes are measured by their fraction rejected. In the example of coffee service, if the average number of spilled coffee cups is one cup per 1000 cups served, then the rejection rate is 0.001, suggesting a rare event, at least statistically. However, this measure should be balanced with known industrial statistics.

The reason for the broader perspective of industrial norms is that although the "rare event" is determined by the frequency of event, engineers are in accordance with the law that proper procedures and personnel training must be in place and in play, which is usually best demonstrated empirically. If so, then the process is in control and the law and engineering can reach an accord on fault in litigation involving rare events.

References

Abraham, B. and Ledolter, J. (1983). *Statistical Methods for Forecasting*. New York: John Wiley & Sons, Inc., p. 225.

Advanced Product Quality Planning (APQP) and Production Part Approval Process (PPAP) (2015). PPT, slide 102. Dublin, Ireland: Eaton Corporation. http://www.eaton.com/Eaton/OurCompany/DoingBusiness/SellingtoUs/SupplierAPQPProcess/. Accessed September 15, 2017.

Aerospace Standard 9100 (2004). *AS9100B, Aerospace Standard: Quality Management Systems—Aerospace—Requirements*. Washington, DC: Society of Automotive Engineer, p. 8.

American Society for Quality (2016) *About ASQ*: "Shewhart…successfully brought together the disciplines of statistics, engineering, and economics and became known as the father of modern quality control." http://asq.org/about-asq/who-we-are/bio_shewhart.html. Accessed July 4, 2016.

ASQ/ANSI/ISO (2015). *ASQ/ANSI/ISO 9001-2015: Quality Management Systems—Requirements*. Milwaukee, WI: American National Standards Institute and the American Society for Quality, p. 17.

Bhote, K. R. (1991). *World Class Quality*. New York: AMACOM, p. 34.

Bowerman, B. L. and O'Connell, R. T. (1979). *Time Series and Forecasting*. Pacific Grove, CA: Duxbury Press, p. 344.

Box, G. E. P. and Jenkins, G. M. (1976a). *Time Series Analysis, Forecasting and Control*. Englewood Cliffs, NJ: Prentice-Hall, p. 49.

Box and Jenkins (1976b), p. 88.

DeRusso, P. M., Roy, R. J., and Close, C. M. (1966). *State Variables for Engineers*. New York: John Wiley & Sons, Inc., p. 503.

Gilchrist, W. (1976). *Statistical Forecasting*. New York: John Wiley & Sons, Inc., p. 246.

Grant, E. L. and Leavenworth, R. S. (1988a). *Statistical Quality Control*. New York: McGraw-Hill, p. 60.

Grant and Leavenworth (1988b), p. 7.

Harry, M. J. (1997). *The Nature of Six Sigma Quality*. Schaumberg, IL: Motorola University Press, p. 17.

Kalman, R. E., Falb, P. L., and Arbib, M. A. (1969). *Topics in Mathematical System Theory*. New York: McGraw-Hill, p. 74.

Lütkepohl, H. (2005). *New Introduction to Multiple Time Series Analysis*. Berlin: Springer-Verlag, p. 25.

Masanao, A. (1987). *State Space Modeling of Time Series*. Berlin: Springer, p. 231.

Miller, L. A. (1970). "Air Pollution Control: 'An Introduction to Process Liability and other Private Actions." *New England Law Review*, vol. 5, p. 172.

Owen, D. G., Montgomery, J. E., and Davis, M. J. (2007a). *Products Liability and Safety: Cases and Materials*, 5th ed. New York: Foundation Press, p. 18.

Owen et al. (2007b), p. 187.

Ross, S. M. (1989). *Introduction to Probability Models*. San Diego, CA: Academic Press, p. 464.

Shewhart, W. A. (1931a). *Economic Control of Quality of Manufactured Product*. Princeton, NJ: Van Nostrand, pp. 301–320.

Shewhart (1931b), p. 146.

Shewhart (1931c), p. 12.

Siljak, D. D. (1969). *Nonlinear Systems*. New York: John Wiley & Sons, Inc., p. 308.

Stimson, W. A. (1994). *Statistical Control of Serially Dependent Processes with Attribute Quality Data*. A dissertation to the University of Virginia. Ann Arbor, MI: UMI Dissertation Services, p. 36.

10

Forensic Issues in Product Reliability

10.1   Background in Product Reliability

Reliability is a property of a product that results from thoughtful consideration and engineering design, and therefore it may be an issue in forensic considerations. Product reliability is defined as the probability that an item will perform a required function without failure under stated conditions for a specified period of time. Because the term "product reliability" is well established in engineering, I use that term, but I ask the reader to understand that the term also applies to service reliability.

Forensic Systems Engineering: Evaluating Operations by Discovery,
First Edition. William A. Stimson.
© 2018 John Wiley & Sons, Inc. Published 2018 by John Wiley & Sons, Inc.

Reliability engineering is a profession in itself, with its own methodologies and mathematics. Much of the evidence in discovery concerning the presence, absence, or degree of product reliability is technical, and forensic systems analysts will not generally be experts in the field. But they should be familiar with some of the characteristics of reliability so that, in exploring the evidence, they may recognize the possibility of misfeasance. At a minimum, the analyst should be able to recognize whether or not reliability studies were ever performed. Therefore, I ask the reader to review Appendix B prior to reading this chapter.

Product reliability may be at issue either because it was required in a contract, or because a given product may involve a risk to the life of the user and reliability would be expected under the law. Under these conditions, there may be a question of fraud or false claims. The forensic analyst, usually untrained in law, must approach fraud indirectly because an assessment cannot determine whether an improper practice is fraud. A determination of fraud is the conclusion of a legal process which must evaluate the elements of both intent and unlawful gain. Until that point, the practice in question is an allegation of misfeasance (United States Department of Justice, 1997).

In manufacturing, fraud is found in an environment of poor business practices that invariably include the production process. I don't know which comes first—the fraud or misfeasance—but the environment is easily identified. Good business practices are hampered, internal controls break down, and failure modes grow. The quality of product deteriorates and is manifested after delivery by poor product reliability. Hence, there is a chain of events: misfeasance, systemic product failure, and product unreliability. In engineering, reliability is defined rigorously and my purpose in this chapter is to show how it relates to the law implicitly or explicitly. The technical evidence of the existence or absence of product reliability may indicate poor business practices and management misfeasance. Such evidence may be the first step in establishing fraud in manufacture.

The arguments in this chapter are based on three premises:

1) Management misfeasance may lead to systemic process failure.
2) Systemic process failure may lead to product unreliability.
3) Therefore, product unreliability may be a key indicator of possible misfeasance.

Management misfeasance creates possible systemic process failure because it departs from good business practices. Correct processes and procedures require good leadership and constant vigilance to achieve and maintain operational effectiveness. Systemic process failure can lead to a large volume of nonconforming product. Such product will lack design quality from the beginning and product reliability will deteriorate as the volume of nonconformity grows (Taguchi et al., 1989). A product without reliability means that reliability was never designed into it to begin with, or that systemic process failure occurred in manufacture. Thus, we are back to the first point—management misfeasance.

An important part of management validation is focused on "reliability"—what the word means in manufacture. Even in industry, many people get it wrong. Reliability is a property of a product or service that, when properly designed in, adds great value to customers, including buyers of volume such as the Federal and state governments, and supply chain producers such as auto makers. Reliability alone provides user safety. Reliability can also increase the value of those products in which the customer expects substantial life, such as automobiles and major appliances.

Investigating product reliability offers attorneys three benefits in litigation. The first point is the connection between systemic process failure and misfeasance and has already been explained. If systemic failure exists, the possibility of establishing management misfeasance is increased and can be a major factor in litigation strategy. Management negligence, indifference, or malfeasance is indefensible.

That a vendor continues to sell and deliver nonconforming product in volume indicates a difference between what the vendor has contracted to do and what is being done or manufactured. It justifies an investigation into the possibility of false claims, which in turn increases the vendor's liability.

In this chapter, I show why product and service reliability are necessary and why they are often so difficult to obtain. I also discuss how reliability and warranty are related, how an expectation for product or service reliability can be justified even in the absence of contractual requirement, and how reliability problems may indicate fraud or result in false claims.

If product reliability is so important to the consumer, why isn't it included in every product or service contract? There are many answers to that, but the main reason is that adding reliability to the product is not cheap. Reliability expands the scope of design and production, adding cost. Reliability testing may be destructive, adding more cost. The usual practice is to offer warranty in lieu of reliability, which in many cases satisfies customer expectations. However, it cannot be repeated too often that warranty and reliability are not the same thing; indeed, they have different purposes.

In many industries, testing is considered a non value-adding activity and may be waived by the contractor when the production schedule gets tight. I discussed, in Chapter 3, the tendency to misread Lean policies in order to justify abandoning tests. For many companies, corporate focus is solely on the bottom line. But such focus can lead to negligence of good business practices, poor product quality and reliability, unacceptable production methods, incorrect processes, and unethical procedures.

10.2 Legal Issues in the Design of Reliability

In regard to processes of production and service, numerous precedents have been established in court that can guide forensic systems analysts in their examination of evidence in discovery. A few of the legal trends most relevant to systems design are as follows:

- Good design practices are viewed favorably by the courts.
- Intended use is incumbent upon contractor and supplier.
- The design process provides a paper trail of evidence of performance.
- Design is an intrinsic phase of manufacturing and service.
- The intended use requirements of ISO 9001 (ANSI/ISO/ASQ, 2015), Clause 8.3, *Design and development*, imply that reliability is a design requirement.

ISO 9001 is internationally recognized as a set of good business practices (including design processes) and will serve as our reference. The phases of good design practice are defined in ISO 9001 (ANSI/ISO/ASQ, 2015) and shown in Column 1 of Table 10.1. The second column lists some of the evidence that forensic investigators might look for in discovery. To aid the forensic investigator, there is a paper trail of evidence in the process of design and development, and each phase of the design process has a trail of proper procedures. In particular, we are interested in whether the design of the product includes its reliability, where appropriate.

The requirement to demonstrate reliability for *intended use* should be stressed. This idea holds suppliers to the same scrutiny as the contractor if they know the intended use of the final product they are contributing to. For example, a weaver of fabric to be used as body armor may be as responsible for product reliability as the producer of the armor.

Table 10.1 Good design practices from ISO 9001.

Design phase	Evidence
Planning for reliability	Customer requirements; historical data; metrics; minutes of meetings; reports
Design inputs	Performance requirements; applicable statutes and regulations; previous design data of similar subject
Design outputs	Critical production data; acceptance test criteria
Design review	Review team; evaluation analysis; review reports
Validation and verification	Intended use requirements; feasibility studies; prototype test results
Change control	Authorization; documentation

10.2.1 Good Design Practices

To my knowledge, no court of law has ruled that companies must register to ISO 9001 or even conform to the Standard unless contractually obliged to do so. But in the chaise longue case discussed in Chapter 8, the defendant chose to settle out of court rather than to risk a trial in which it would be clear to the jury that the defendant company did not conform to what the whole world thinks is the correct way to do something (Broomfield, 1996).

Further demonstration of the legal importance of good design practices is found in a case also discussed in Chapter 8, where judge Thomas Eisele, of the US District Court of the Eastern District of Arkansas ruled that, pursuant to Federal Rules of Civil Procedure 401 and 402, "all 'relevant' evidence is admissible and evidence is 'relevant' if it tends to make a fact that is consequential to determining that an action is more or less probable than it would be without the evidence" (Kolka, 2009).

Judge Eisele further noted that "the US Eighth Circuit Court has consistently held that compliance or noncompliance with ANSI standards is admissible in strict liability and negligence cases in order to prove a design defect." As ANSI standards of quality management are recognized internationally as good business practices, the point is made.

10.2.2 Design Is Intrinsic to Manufacturing and Service

Every product or service begins with a concept. It is axiomatic that design is intrinsic to manufacturing and service. However, every phase of manufacturing can be outsourced and ISO 9000 recognizes that some companies do not do design. If a company does not do design, it may still get certified to ISO 9001 by meeting the requirements of ISO 9001:2015, Clause 4.3 *Determining the scope of the QMS.* This clause offers conformity to the Standard if particular requirements (e.g., design) do not affect the organization's ability to ensure the conformity of its products and services.

Hence, a company may outsource a design or procure one through purchase. But the company must still meet the requirement of Clause 8.4, *Control of externally provided processes, products and services,* that externally provided processes, products, and services conform to requirements. Moreover, *FAR 52-246-2(b), Inspection of Supplies* (Federal Acquisition Regulations, 1996), holds the contractor to the requirements of Clause 8.3 *Design and development of products and services* if such is in the contract.

10.2.3 Intended Use

Intended use logically implies future use and therefore that product quality is expected to have a duration, hence reliability. Responsibility for quality, and by extension reliability, is continuous through the supply chain and extends to the intended use of product. This responsibility cannot be delegated and depending on the contract it is supported by various authorities, listed in the following:

- *FAR 52-246-2(b), Inspection of Supplies* (Federal Acquisition Regulations, 1996), which admits supplier review, does not relieve the contractor of the obligations under the contract.
- *FAR 46-405(a), Subcontracts* (Federal Acquisition Regulations, 1995), recognizes contract QA on sub-contracted supplies or services, but does not relieve the prime contractor of contract responsibilities.
- *ISO 9001:2015,* Clause 8.2.3, *Review of the requirements for products and services,* requires the organization to review the requirements not stated by the customer, but necessary for the specified or intended use, where known.
- Clause 8.3.4, *Design and development controls,* requires controls to ensure that validation activities are conducted to ensure that resulting products and services meet the requirements for specified application or intended use.
- Clause 8.3.5, *Design and development outputs,* requires the organization to specify the characteristics of the products and services that are essential for their intended purpose and their safe and proper provision.

Whether the design is achieved in-house or through an external source, the company must meet the intended use requirement. They must, before delivery, verify the ability of their product to perform its intended use. If its design is deficient to this end and they sell the product anyway, then they are in violation of Clause 8.2.2, *Determining the requirements for products and services*; Clause 8.3.4, *Design and development controls*; and Clause 9.1, *Monitoring, measurement, analysis and evaluation.*

10.2.4 Paper Trail of Evidence

The paper trail can be derived from the schematic of design and development shown in Figure A.1:

- Documented customer requirements
- Specifications
- Drawings
- Models, prototypes, feasibility tests, and demonstrations
- Analytical data from design reviews, that is, cause and effect diagrams
- Test documentation: plans, verification, validation, accelerated test criteria
- E-mail concerning any of the above

Failure modes and effects analysis (FMEA) and risk management procedures are useful in the design phase of a product to anticipate where failure might occur. They may also be used after in-service failure to trace the cause. A defendant's use of FMEA and risk management to determine reliability is the first evidence of reliability design. Conversely, its absence is a flag for concern.

For example, in one case I worked on, the premature failure of a product was traced to inadequate production and incorrect testing procedures, both of which lacked risk analysis in their development. In another case, product quality was tested and the results were very good. However, product reliability was never tested anywhere in the supply chain and it quickly became clear that the product had no reliability at all. It began to fail contract requirements almost as quickly as it was delivered.

10.2.5 Reliability Is an Implied Design Requirement

Reliability is not mentioned in ISO 9001, but it can be inferred from careful reading of Clause 8.5.1, *Control of production and service provision*, which requires controlled conditions including the validation and periodic revalidation of the ability to achieve planned results of processes where the resulting output cannot be verified by subsequent monitoring or measurement.

By definition, product reliability in the field can be determined only by the product being in use. An expected value for reliability can be determined in the laboratory, but this estimate must always be verified by field data.

The sole argument is this: what is meant by "output" in the requirement of the clause cited above? Obviously, the outputs of the processes are products and services and verification is required of their key characteristics. There are two: quality and reliability as indicated by the "intended use" requirements cited earlier. Thus, the production processes are validated when the quality and reliability of their products are demonstrated in statistical conformance to specifications.

10.3 Legal Issues in Measuring Reliability

As with product design, the measuring of product or service characteristics also has certain legal issues:

- All product lines are subject to phases of failure. A stable production system provides product with a period of random failure, whose life statistics can be measured.
- With Weibull analysis, product mean time to failure (MTTF) can be estimated from this stable period.
- As with all statistical measurement, the more failure data the better.
- The measuring process provides a paper trail of evidence of verification and validation.

10.3.1 Failure Modes

Infant mortality is a fact of life and wear-out is inevitable. Between those two failure modes there can exist a stable mode in which product or service operates as designed and in which reliability can be measured. This stable mode is provided by a stable production system and good design.

In the stable mode, failure will occur randomly. Product life is stochastic and there is always common cause of failure. Estimation can be made of the average life expectancy of products in this stable mode. This estimation, be it MTTF, mean time between failure (MTBF), or B50, is what you are paying for when you buy reliability.

However, special cause can result in systemic failure, which is almost always due to mismanagement. If the product line has attained the wear-out mode prematurely, the source of special cause must be determined.

10.3.2   Estimation of MTTF

If the industrial process is stable, then product failure has a distribution from which failure rates and mean life can be measured. This is achieved with a technique called Weibull analysis, discussed in detail in Appendix B, which matches failure data to distributions that represent the failure rates of products under study. Given a set of failure data, then the underlying distribution of the failure mode can be determined. Weibull analysis has become part and parcel of reliability engineering. The estimates are effective in law and contractually valid. For example, many US Navy contracts include reliability in their specifications.

A Bayesian technique is one in which estimates of parameters can be made based on previous information. (Thomas Bayes was an English mathematician of the eighteenth century.) Combined with Weibull analysis, Bayesian methods offer a powerful tool for making estimates of product life. Equation 10.1 shows an example of a Bayesian method.

Suppose we know the shape of the failure distribution of a given product from previous information. Say that the shape is described by the parameter $\beta = 1.5$. Then we can determine the value of η, the characteristic life of the product, with a few failure data points as shown: 820, 1130, 1425, 1543, and 1708 hours. In Equation 10.1, η represents the expected life of the product.

$$\eta = \left[\frac{(820)^{1.5} + (1130)^{1.5} + (1425)^{1.5} + (1543)^{1.5} + (1708)^{1.5}}{2.303} \right]^{1/1.5} \tag{10.1}$$

The division by 2.303 converts the natural log value to a base-10 log value (Zaretsky et al., 2003). In this case, the expected life of the product is $\eta = 2254$ hours.

You may ask, "Why should I be interested in this equation? It's pretty esoteric."

There are two reasons. First, evidence of Bayesian methods may be found in discovery and forensic systems experts should be able to recognize them.

I know of a trial attorney, in his youth an engineering student, who did exactly that and the evidence turned out to be significant support for false claims. Forensic systems analysts may not always be able to interpret technical evidence, but they should be able to recognize its nature and determine whether additional technical expertise is required.

Second, the method can be used as a shortcut to predict the MTTF of a newly designed product, assuming that the β estimate has not changed from a previous and similar design. However, this is not a safe assumption in new design and you are justified in challenging it if field data from the new design do not support the estimate. In this event, it may well mean that the design or the design process was wrong. The attorney challenged this assumption in his case and, furthermore, found defendant engineers who challenged it also, weakening the defense's position.

10.3.3 The More Failure Data the Better

There is an axiom in geometry that says the shortest distance between two points is a straight line. A corollary to this axiom is that two points can only define a straight line. I have said that if failure data follow a straight line on log paper, then the data can be analyzed for Weibull distribution. But if the data form a curved line, then Weibull analysis is not appropriate.

Therefore, two points of failure data are inadequate, for the underlying data may be curved. So how many data points are needed? That depends upon how much is known about the designed product. If the new design is quite similar to a previous model with well-known reliability data, then perhaps only a few data points are required to estimate the failure rate of the new design. But even in this case, caution suggests that the more failure data the better for a valid Weibull analysis.

Some companies will claim that they have such profound knowledge of the system under study that they can do Weibull analysis on two data points and even one data point. The forensic analyst should be skeptical of such a claim. It may be true if the system has not changed, but in a new design or in redesign, the system may well have changed and the change may go undetected. It is unsafe to assume that a system remains the same after a change in design is made, and additional failure data points may be required to achieve a valid analysis.

10.3.4 The Paper Trail of Reliability Measurement

The paper trail generated in reliability measurement includes the following:

- Weibull charts and graphs
- Specifications of product life: MTTF, MTBF, and failure rate
- Records of analysis and measurement results

- Minutes of measurement reviews
- Test documentation: plans, verification, validation, and accelerated test criteria
- E-mail concerning any of the above

The purpose of measurement is to verify and validate product design, development, and fabrication, including its reliability. The results of reliability measurement tell you whether the design was successful or not and if not, and you paid for a successful design, then you must continue your inquiry into fraud and false claims.

I repeat from Chapter 4 the quote from attorney Jim Kolka (2004): "The good news is that quality leaves a paper trail. And the bad news is that quality leaves a paper trail." It means that if you are doing the right things, then your paper trail will help you if you must defend yourself in court. If you are doing the wrong things, then the paper trail will help the plaintiff. As reliability can be defined as quality over time, the adage pertains to reliability also.

10.4 Legal Issues in Testing for Reliability

All phases of the design and production of product reliability provide evidence to pursue in seeking redress in litigation. Evidence one way or the other will abound. Evidence of poor reliability provides an avenue for damages. Evidence of good reliability saves wasted time in fruitless pursuit. In the case of life testing of product or service, the legal issues that arise might be resolved with the following evidence:

- Defined and documented life test procedures must be in accord with accepted standards.
- Life test records and reports; MTTF and MTBF must be in accord with specifications.
- Test procedures, specification requirements, and test objectives are defined á priori.
- À posteriori changes to test procedures and results must be justified.

10.4.1 Defined and Documented Life Test Procedures

The forensic systems analyst will verify the existence and use of documented procedures for life testing used to establish product reliability. Such definitions and procedures should be in accord with national or industrial standards. Available documentation will demonstrate whether the performer was at least capable of estimating product reliability and indicates a preliminary effort.

10.4.2   Life Test Records and Reports

The forensic systems analyst will verify the existence of records and reports of life testing, in which metrics have been defined and relevant measurements made. The measurements should be in accord with the test and/or contract specifications. These records provide evidence that the defendant did indeed conduct life testing of its products. Life testing records will include Weibull analysis and it is well worthwhile to recognize a few of the Weibull procedures.

Weibull analysis is basic to reliability design and verification. There should be evidence of it in the design and development activities described in discovery. Even if you have no background in reliability, with a little familiarity of the subject you can recognize evidence of reliability studies. In examining the evidence in discovery, you can make three important assessments in regard to reliability: (i) If there is no evidence of reliability studies in discovery, it is highly likely that none have been done. (ii) If there is such evidence, then an estimate of the product life, MTTF, should be available. (iii) If the estimated MTTF is not supported by field data, it may be well to suggest to counsel that a reliability expert be brought in to help determine the adequacy of the design procedures in litigation. This can be a critical issue because of the complexity of reliability analysis. Some analyses may appear satisfactory at a cursory level, but may be in fact inadequate.

10.4.3   Test Procedures

The forensic systems analyst will verify that the procedures and specifications to be used in a test and the objectives of the test are defined *before* the test and analysis are begun. The correct sequence of test actions ensures objectivity in going from test design to test result. An incorrect sequence, say changing test specifications after the test, suggests writing a test to achieve desired results, a serious violation of test integrity.

The propriety in changing a test procedure is rather difficult to resolve. Certainly, it is unethical to change the resulting test data in order to conform to objectives. Nor is it ethical to change a test procedure and rerun a test in order to get favorable results. However, following the design of a new product, a test may indicate that the product was incapable of achieving specified results. Then either the product or the test itself was inadequately designed. This can happen even in the best environment and as long as corrections and verification are recorded and completed prior to sale, no harm is done. Nevertheless, one can say generally that á posteriori changes to test procedures must be justified. In addition, the contract involved in litigation may require that the customer be informed of such changes.

10.5   When Product Reliability Is not in the Contract

There are two related ideas in this discussion whose names may cause some confusion to the reader because of their similarity in sound: product *liability* and product *reliability*. They are not related in definition but in legal interaction. Reliability has to do with the quality of a product over time; liability has to do with harm caused by a product. If that harm occurs because of the premature failure of a product, then product liability applies and perhaps product reliability also.

10.5.1   Product Liability

If the requirement for product reliability is in the contract, then contract law obtains and the pursuit of litigation is clear. But there may be cases in which reliability of product or service is incumbent upon a performer whether or not it is in the contract. Some issues that suggest the provision of product reliability would be (i) legal trends; (ii) safety of users; (iii) industry code of practice; (iv) due diligence; and (v) catastrophic failure.

The history of product liability litigation has evolved from the burden being borne by the plaintiff, to the notion of strict liability in which the performer is at risk even without negligence or warranty breach. Tort law, too, has something to say about this and is used to force a remedy for product liability cases where there is (i) negligence, or failure to maintain standards; (ii) breach of warranty, or failure to provide a product that meets the warranty; and (iii) strict liability, in which manufacturers are held liable even when without fault (Boehm & Ulmer, 2009a).

Courts now look beyond the defect of a product to a company's QA procedures to determine their ability to prevent and control risks. This may include an assessment of the company's effectiveness in all processes: design, marketing, manufacturing, shipping, service, and supplier selection. Existing quality systems provide a formal mechanism for most of the facets of due diligence demanded by the courts.

Exposure to liability exists throughout the product life cycle from design, through production, warranty, in-service to end of life. Product designers must consider the ultimate use of the product for safe use. In production, the product must meet such specifications that include safety features against hazards that cannot be designed out, such as string trimmers and weed whackers. In marketing, warranty and advertising claims must accurately reflect product life cycle, safety, and performance expectations.

The requirement to provide seat belts in autos is an example of a specific legal requirement for reliability. If seat belt use is mandatory, then a reasonable person would assume that the belt is reliable. It seems unlikely that the courts would uphold the argument of a seat belt producer that it did not need to provide product reliability.

In general, the safety of users can mandate reliability whether or not there is a specific legal requirement. The Ford/Firestone tire failures of 2001 are a good example of a change in design that resulted in a reduction in reliability, causing great bodily harm. The manufacturers involved faced hundreds of millions of dollars in damages and billions in recall costs (Stimson, 2001).

A given industry may have a code requiring reliability and the manufacturer who does not follow suit will have a weak defensive position in liability. In the spirit of due diligence, it would be wise for designers to incorporate aspects of a code appropriate to their product and industry. *Due diligence* refers to an established legal view of a measure of prudence, responsibility, and diligence that is expected from a manufacturer or provider of services under the circumstances (WebFinance, Inc., 2016).

10.5.2 ISO 9001 and FAR

Beyond the possible legal issues, a good case can be made for damages for failed product reliability when the contractor is certified to ISO 9001. This pursuit is important because many performers are certified to the Standard and often qualify to bid on requests for proposal because of their certification. If a company wins a bid when ISO 9001 was a requirement for bidding, then the certification is material.

Companies that are ISO 9001 certified may have a stronger responsibility than they realize to design reliability into their products and services and there are two arguments to justify this assertion: (i) design for reliability is implicit in ISO 9000, recognized all over the world as a standard of good business practices and (ii) reliability is a good business practice and its omission demonstrates a lesser standard of care.

In regard to the first argument, reliability requirements are not explicit in ISO 9001 and you won't find the word there. But both ISO 9000 controls and Federal Acquisition Regulations suggest reliability. A performer would be in a poor defensive position if it argued that it had no obligation under the Standard to supply reliability of product.

In regard to the second argument, ISO 9001 is an internationally recognized set of good business practices that are known to either contribute to the quality of product and service or whose absence invites quality failure. ISO 9001 does this comprehensively, covering activity in four general categories: management, resources, operations, and improvement. Each requirement within these categories serves as an instruction, teaching the reader the elements necessary to achieving the requirement. For example, Clause 8.5.3, *Preventive action*, requires actions that are essentially product FMEA, a basic reliability analytical tool.

Seen in its entirety, the ISO 9000 standard places a burden on its certified performers to provide reliability in their products and services, or risk litigation and damages.

Traditionally, we assume that if a product is offered for sale, it should be fit for intended use (Boehm & Ulmer, 2009b). This is an important assumption in ISO 9001, which has several requirements for the intended use of product.

The catastrophic failure of a product or system within its specified environment is completely unacceptable. Therefore, it is important that a FMEA be carried out in the design phase of a product. The failure of the booster seals in NASA's Challenger spacecraft comes to mind. Given the testimony of Elon Musk (2003), if NASA had been a private company, the liability could conceivably have put it out of business.

In sum, good business practices are expected of a performer, at least to a certain measure. The expected measure is well defined when the performing company is certified to a recognized standard, but legal decisions have imposed expectations that may include product reliability even upon noncertified performers.

10.6 Warranty and Reliability

For purposes of this book, I define a warranty as an obligation that a product or service sold is as factually stated or legally implied by the seller and provides for a specified remedy in the event that the product or service fails to meet the warranty. The key word in this definition is *fail.* Both warranty and reliability are keyed to product failure and because warranty implies a period of acceptable performance, warranty and reliability are often thought to be equivalent. They are related—at least they should be (Wiener, 1985)—but they are not the same thing nor are they equivalent. Granted that warranty and reliability may be complementary and both useful to provide a complete service to the customer, they are nevertheless very different in nature, purpose, and definition.

Reliability is a characteristic of a product that is designed to provide a given life span. As no design or production system is perfect, the products will realize a varying life span. Reliability is the average value of product life, which means that about half the products will fail earlier than the average life and about half will fail later.

The purpose of a warranty is to provide a defined compensation in the event of product failure. Few companies can afford to warrant half the product line that is expected to fail prior to the average of the product population, so the warranty period will be considerably less than the specified average life.

Both vendor and customer benefit from warranty. The vendor gets a contractual agreement established on sale of product that defines buyer responsibility, warranty limitations, and vendor liability. The warranty also serves as a marketing tool and, at least in principle, provides a measure of product reliability. The customer's benefit is an implied durability and an agreed compensation in case of product failure within the warranty period.

However, warranty and reliability must not be confused. Warranty, like durability, is not an engineering term and means only what the performer wants it to mean. In a perfect world, the warranty of a product would be based on reliability studies, but sometimes this is not the case and cannot be the case if reliability studies have never been made. Sometimes a warranty is determined simply as a trade-off between warranty reserves and the probability of payout. On a few occasions a product warranty may be based on nothing more than what others are doing. "That's what our competition is offering, so we will offer it too."

A grave problem exists when the relationship between warranty and reliability is vague or even nonexistent. Every productive process suffers variation and even in a good production system there will be some unacceptable deviation, with resulting producer and consumer risks. In designing the production system, a producer may choose a lower producer risk and warranty reserve, accepting the higher consumer risk in a trade-off against the cost of reliability. It is a pure business decision. Sometimes a warranty is offered because to do so is an industry standard. Sometimes warranties are simply created out of thin air, as when reliability was never designed into the product.

If warranty is in the contract, then failure data should be available and provide real-time evidence of product reliability, if it exists. The data verify or deny the initial reliability estimates. The relationship between reliability and warranty is symbiotic in a company with effective systems of design and production that include product reliability.

Some attorneys are comfortable with warranty, however minimal its justification, because it is a contract commitment and default can be easily litigated. As an engineer, I cannot argue with this strategy but hasten to add that warranty may protect the buyer, but it does absolutely nothing for the user. If a loved one's seat belt fails and that person is killed or permanently injured, what good is the warranty?

Therefore, if product durability is an implied factor in an investigation, then inquiry into product reliability is appropriate whether or not it is in the contract. You must determine the basis of any contracted warranty because there may be possible fraud. The occurrence of failures is statistical and manufacturers know this. They may weigh producer risk against consumer risk and cover the difference with warranty, gambling that statistics are in their favor. However, if there is systemic failure, statistics go out the window and buyer and seller become gamblers. Warranty unsupported by reliability data is technically weak and can be challenged.

References

ANSI/ISO/ASQ (2015). *ANSI/ISO/ASQ Q9001-2015: Quality Management System—Requirements*. Milwaukee, WI: American National Standards Institute and the American Society for Quality.

Boehm, T. C. and Ulmer, J. M. (2009a). "Product Liability Beyond Loss Control—An Argument for Quality Assurance." *Quality Management Journal*, vol. 15, pp. 7–19.

Boehm and Ulmer (2009b), p. 10.

Broomfield, J. R. (1996). "Lawyers Wise Up to ISO 9000." *Compliance Engineering*, vol. 13, no. 1, pp. 32–39.

Federal Acquisition Regulations (1995). *FAR 46-405(a), Subcontracts.* http://farsite.hill.af.mil/reghtml/regs/far2afmcfars/fardfars/far/46.htm#P185_28365. Accessed September 15, 2017.

Federal Acquisition Regulations (1996). *FAR 52-246-2(b), Inspection of Supplies.* http://farsite.hill.af.mil/reghtml/regs/far2afmcfars/fardfars/far/52_246. htm#P8_708. Accessed September 15, 2017.

Kolka, J. (2004). *ISO 9000: A Legal Perspective.* Milwaukee, WI: ASQ Quality Press, p. 45.

Kolka, J. (2009). *ISO 9001 Lawsuits & Document Control #2.* http://jameskolka. typepad.com/international_regulatory_/2009/03/index.html. Accessed September 20, 2017.

Musk, E. (2003). Testimony before a Joint Hearing on Commercial Human Spaceflight: U.S. Senate Science, Commerce and Technology Subcommittee on Space; U.S. House of Representatives Subcommittee on Space and Aeronautics, July 24, 2003. The witness testified that unlimited liability would not only prevent a company from offering a service such as space flight, but it would kill the entire industry. Mr. Musk is CEO of SpaceX Corp.

Stimson, W. A. (2001). *Internal Quality Auditing*, 2nd ed. Chico, CA: Paton Press, p. 265.

Taguchi, G., Elsayed, E. A., and Hsiang, T. (1989). *Quality Engineering in Production Systems.* New York: McGraw-Hill, p. 14.

United States Department of Justice (1997) *United States Attorneys Manual, Title 9.* https://www.justice.gov/usam/criminal-resource-manual-949-proof-fraudulent-intent/. Accessed September 15, 2017.

WebFinance, Inc. (2016). http://www.businessdictionary.com. Accessed July 7, 2016.

Wiener, J. L. (1985). Are Warranties Accurate Signals of Product Reliability? *Journal of Consumer Research*, vol. 12, September, p. 245.

Zaretsky, E. V., Hendricks, R. C., and Soditus, S. M. (2003). *Effect of Individual Component Life Distribution on Engine Life Distribution.* NASA TM-2003-212532. Washington, DC: National Aeronautics and Space Administration, p. 18. https://ntrs.nasa.gov/archive/nasa/casi.ntrs.nasa. gov/20030093640.pdf. Accessed September 15, 2017.

11

Forensic View of Internal Control

All conservatism is based upon the idea that if you leave things alone, you leave them as they are. But you do not. If you leave a thing alone, you leave it to a torrent of change.

(G. K. Chesterton, 1908)

A process is defined internationally as a set of interacting activities that turn inputs into outputs (ISO 9000, ANSI/ISO/ASQ, 2005). But in Chapter 9 we saw that there is no such thing as a perfect process and there will always be variation in the key characteristics of the output of any process as a result of special and common causes. Some of that variation will not conform to requirements. We need a way to ensure that stable and capable processes remain so through self-correction. Stability provides consistency in output characteristics. Capability provides cost-effective results. System self-correction reduces or eliminates nonconformity in product or service through a system of controls.

Forensic Systems Engineering: Evaluating Operations by Discovery,
First Edition. William A. Stimson.
© 2018 John Wiley & Sons, Inc. Published 2018 by John Wiley & Sons, Inc.

11.1 Internal Controls

Controls are essential to diverse professions, each with its own terms of art. For example, internal controls are framed by the law, which imposes constraints on financial dynamics and procedures in order to ensure market transparency. "Internal control" comes from the financial audit world. That the term comes from the audit profession does not prevent its use in systems engineering. Forensic systems analysts are at the same time auditors of operations, and effective audit techniques are appropriate to their work. As a systems engineer, I have been using the term for the past 10 years because I admire the formality and approach of financial audits and because the term fits in well with the notion of a system (Stimson, 2005). In this book, if a control is intrinsic to the system it controls, then it is an internal control.

11.1.1 Purpose of Control

Correction is the purpose of control. Real-time control minimizes the buildup of errant behavior by providing timely correction. A control is simply an arrangement that, together with the process, ensures a desired result. The control might be a hardware device, software program, or even a management decision system. To evaluate the effectiveness of an internal control, one must understand both its nature and how it fits into the process flow.

Business, too, favors internal controls. According to the Committee of Sponsoring Organizations of the Treadway Commission (2013), "senior executives have long sought ways to better control the enterprises they run. Internal controls are put in place to keep the company on course toward profitability goals and achievement of its mission, and to minimize surprises along the way. They enable management to deal with rapidly changing economic and competitive environments, shifting customer demands and priorities, and restructuring for future growth. Internal controls promote efficiency, reduce risk of asset loss, and help to ensure the reliability of financial statements. They also ensure compliance with laws and regulations."

Internal controls are mandated by the Sarbanes–Oxley Act of 2002 to ensure the transparency of financial audits (United States Congress, 2002). However, careful reading of the Treadway statement reveals that the conversation goes beyond financial operations and applies to corporate operations across the board (Stimson, 2012).

In engineering, control is used in processes of any type to ensure that the actual output agrees with the desired output. Obviously, this calls for a comparison of the two and suggests the solution: the use of an arrangement in which the output of a process is sampled and compared to that required. A difference generates a corrective signal from the controlling device that is fed back into the process in such way as to realign the actual output with that desired. Such a system is called a closed-loop control system.

11.1.2 Control Defined

Engineers have various ways to define control. One will say that a control is a device to modulate the behavior of a process in some desirable way. Another will say that a control relates cause and effect through a performance function. If the process is stochastic, one might say that a control enables us to predict, at least approximately, the probability that the process output characteristics will fall within given limits.

Although in this chapter we shall discuss forms of control in engineering terms, to the forensic systems analyst the most appropriate definition of internal control is that given in Chapter 5 and repeated here for convenience: An internal control is a process designed to provide reasonable assurance regarding the effectiveness of operations; reliable records and reports; and compliance with regulations. This definition is also used by the Securities and Exchange Commission (2003).

This definition satisfies the work of the forensic systems analyst because it is process oriented and conforms to traditional engineering notions of system effectiveness and stability. Moreover, it responds to legal requirements, as must operational activities that appear in a contract. In forensics, "regulations" include appropriate laws and standards.

Therefore, the task of the forensic systems analyst is to determine whether the internal controls in litigation are effective, reliable, and compliant. An adverse decision by the analyst will be challenged, so it is incumbent upon that person to know why and how controls work. In this chapter, we shall examine how controls are designed and implemented to affect desired outcomes. The design establishes control stability and capability. The implementation establishes control integrity—whether it is doing what it is supposed to be doing.

11.1.3 Control Elements in Operations

Figure 11.1 shows a general diagram indicating the resources necessary to processes of production and service. These resources indicate the nature of the controls needed. The physical equipment are those hardware and software

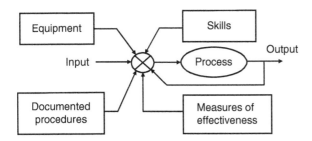

Figure 11.1 "Turtle diagram" of a generalized process.

devices necessary to the process, such as monitoring and recording devices, embedded computers and associated software, position and rate controls, counters, and cameras. The documented procedures are those that describe the process set up and the initial conditions of process and unit to be processed, rules and regulations, instructions on process operation, technical manuals, schematics, and associated information needed for proper operation. Skills include those necessary or appropriate to the process operation and maintenance, and the operators with those skills. The measures of effectiveness (MOE) include those metrics necessary to verify and validate both the process and the product to ensure the attainment of process objectives and contract requirements.

In examining the effectiveness of a process, forensic systems analysts will look for evidence of the presence or absence of proper, competent and timely skills, procedures, MOE, and proper procedures and instruments for verification and validation. By proper is meant those metrics, procedures, and devices that are accepted industry wide and by the customer, and that maintain test integrity. The analyst will further consider whether the process inputs were acceptable—those described in the company flow chart and in the formally approved procedures.

A process is said to be robust if it is self-controlled to maintain its effectiveness when subject to disturbances—those undesired human and natural forces that are external to the designed process, but act upon it. These disturbances cause variation in the process output and a correction must be made to return the system to its designed stability and capability. Another name used for such disturbances is "special cause" and they must be identified and eliminated.

11.2 Control Stability

An effective control is integrated with the dynamics of the system that it controls and so analysis of system stability includes that of its control. Control stability is determined in the design phase of a process by the use of various mathematical models, usually under the assumption that the system is linear. Although some systems may be nonlinear, they can be approximated by linear systems over a limited range, thereby easing the design problem. A system is linear if it can be described as

$$k_1 f_1(t) + k_2 f_2(t) \rightarrow k_1 y_1(t) + k_2 y_2(t), \tag{11.1}$$

where $f(t)$ is the forcing function, or system input, and $y(t)$ is the system output. The symbol $\rightarrow$ means y responds to $f(t)$, and the coefficients, k_i, are often assumed to be time invariant, as most systems observed in practice belong to this category (Lahti, 1965a).

The principle of superposition indicated by Equation 11.1 means that a complex forcing function input that can be broken into a sum of simpler inputs will yield the same response. Generally, any forcing function, $f(t)$, encountered in practice can be expressed as a discrete or continuous sum of exponential functions, e^{st} (Lahti, 1965b). The response of a linear time invariant system to an exponential function is also an exponential function. This relationship is useful because a large variety of dynamic behaviors can be expressed as exponential functions.

The response of a linear system to any input has two components. One is the natural response of the system itself and depends entirely upon its design. This component is called the transient response and will die out if the system is stable. The second component is called the steady state response and depends on the nature of the input. It too may die out according to the type of input. If the transient response dies out more quickly, the steady state response is said to dominate. In steady state dominance, the system will respond to an exponential forcing function with an exponential response of the same frequency.

A dynamic system can be described by differential equations if it operates in continuous time or by difference equations if its operational time is discrete. Either view may be taken, depending on how the operation is to be characterized. For example, production and service systems are discrete in the sense that the input is often periodic—parts arrival, customers in queue, decisions—hence difference equations are useful and appropriate. This approach is used in Chapter 15. But the systems can also be viewed as piecewise continuous, and differential equations can be used to model them. In either case, linear equations with constant coefficients can be used because operations tend to follow the principles of superposition and homogeneity (DeRusso et al., 1966; Šiljak, 1969).

11.2.1 Model of a Continuous System

If a process is modeled as a piecewise continuous operation, then linear differential equations are appropriate. A typical model might be

$$a_n \frac{d^n y}{dt^n} + a_{n-1} \frac{d^{n-1} y}{dt^{n-1}} + \cdots + a_1 \frac{dy}{dt} + a_0 y = f(t), \tag{11.2}$$

where y represents the output variable and $f(t)$ is the forcing function, or input variable. The differential equation represents the dynamic response of the process to the input, $f(t)$.

In general, nth order differential equations can be difficult to solve for large values of n, and in control system analysis it is often convenient to convert them into algebraic equations by use of the La Place transform.

The operator of a La Place transform is the letter s, a complex number ($s = \sigma \pm j\omega$), so Equation 11.2 is transformed by

$$Y(s) = \int_0^\infty y(t)e^{-st}\,dt \tag{11.3}$$

which when applied to Equation 11.2 yields

$$\left(a_n s^n + a_{n-1} s^{n-1} + \cdots + a_1 s + a_0\right)Y(s) = F(s). \tag{11.4}$$

Equation 11.4 is an algebraic expression in s and is presumably easier to solve than in its differential form. Formally, the lower limit of the La Place transform is $-\infty$. However, for our purpose time begins at zero because the factor e^{-st} grows without bound for large negative values of time if σ is a positive number.

When there is no input to the system, $F(s) = 0$, and Equation 11.4 is then called the characteristic equation of the process, because the roots of the polynomial indicate the inherent stability of the process. There may be both real and complex roots of the equation, but the process is stable if and only if the real part of all its roots is in the left half of the complex plane (negative real values). Briefly, a stable process is one that responds with a bounded output when subject to a bounded input.

When $f(t)$ is a dynamic force, it, too, will have a La Place transform, $F(S)$, expressed as a polynomial in s with constant coefficients, b_j. Then Equation 11.4 is expressed as

$$\left(a_n s^n + a_{n-1} s^{n-1} + \cdots + a_1 s + a_0\right)Y(s) = \left(b_m s^m + b_{m-1} s^{m-1} + \cdots + b_1 s + b_0\right)F(s). \tag{11.5}$$

Process stability is designed by means of mathematical models of the process and the selection of critical characteristics such that all roots of the unforced system have real parts in the left half of the s-domain, that is, the complex plane. It is not likely that discovery would yield enough data for the forensic analyst to reverse engineer an operational process in order to determine its stability and such an effort would probably be unnecessary. Presumably, there will be control charts that indicate whether the designed process was at one time stable. The most important question is whether the process was stable during the period in litigation. We address that issue in Chapters 14 and 15 because very often there are little stability data available in discovery other than that indicated by nonconforming product.

11.2.2 Transfer Functions

Engineers build from mathematical models. In its design phase, a process often begins as an nth order polynomial in the s-plane as previously described. Parametric values are selected to ensure process stability and at some point

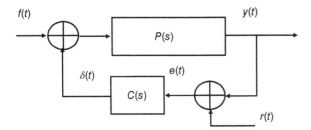

Figure 11.2 A closed-loop control system with corrective path.

inputs are applied and outputs measured to verify the design. The analysis of process operation subject to acceptable inputs begins with the concept of a *transfer function,* which is a representation of the relation between the input and the output of a linear time invariant system. This relationship is most useful to forensic systems analysts because it gets to the heart of the matter—what a process is supposed to do.

A transfer function is simply the ratio of the output signal to the input signal when all initial conditions are zero (Murphy, 1965a). However, they can also include the control dynamics of the process and we shall discuss various implementations in the next section, but at this point we shall examine a general graphic of the transfer function of a closed-loop control system.

Figure 11.2 graphically depicts a closed-loop control system and is identical in form to Figure 7.1, but with detailed notation of the corrective path. $P(s)$ in the top block is the polynomial that represents the forward transfer function of the process. It is also called the open-loop transfer function because if the feedback were removed, then $P(s)$ would be the sole transfer dynamic, as we shall discuss in a moment. $C(s)$ represents the control dynamics and is the feedback transfer function, providing correction for errors in the output. The system input is the forcing function $f(t)$, and $y(t)$ represents the system output. The reference signal, $r(t)$, often called the *set point,* is compared to the output and the difference $e(t)$ represents the output error. Finally, $\delta(t)$ is the correction signal to the process.

In this book, $f(t)$ is often a step input, the type of input most used in forensic systems. A step input is one that occurs almost instantaneously and does not die out, but it maintains some constant amplitude. An operational example might be the addition of a second shift to the production line, or the issue of a new policy, or a new piece of equipment.

In the feedback, $\delta(t)$ is the correction signal from the control. System inputs and outputs may be vector quantities, but as explained in Chapter 15, operations systems in this book are modeled with single variable inputs and outputs, this being the general interest in forensic systems examination, focusing on a single disturbance. The transfer function of a single input, single output nth order differential equation is a proper rational function of degree no greater than n (Luenberger, 1979).

The transfer function of the closed-loop control system can be found from Figure 11.2:

$$Y(s) = P(s)\left[F(s) - C(s)Y(s)\right], \tag{11.6}$$

where $Y(S) = P(S)E(S)$. Then the closed-loop transfer function is the ratio of $Y(s)/F(s)$:

$$\frac{Y(S)}{F(S)} = \frac{P(S)}{1 + P(S)C(S)}. \tag{11.7}$$

The right-hand denominator of Equation 11.7 is the characteristic equation of the controlled system and a proper rational function when reduced. If all of the roots of the closed-loop characteristic equation have negative real parts, then the process is stable and will remain stable when subject to bounded inputs (Murphy, 1965b).

If there were no feedback loop, or if the feedback loop were opened, then we would have an open-loop control system. The open-loop transfer function can be quickly inferred from Equation 11.7 and is simply the forward transfer function, $P(s)$, when $C(s) = 0$. If the process is designed and implemented for stability, then in principle it is stable and should respond with a bounded output when subject to a bounded input.

Open-loop control systems are anathema to control engineers because they cannot provide correction. (The reader may find it interesting that in the days when Wernher Von Braun and his team ran NASA, they would not accept studies of process dynamics run on a digital computer, which was quite frustrating to us—IBM engineers. The reason for their reluctance was that during its duty cycle, a digital computer has open-loop moments. In those days, these moments could be measured in milliseconds—too long, in the German view. Control networks were designed using small-angle approximations. Without control, the Saturn vehicle pointing errors could exceed small angles within a few milliseconds, rendering network solutions invalid. As a result, all studies were done on analog computers, whose medium is the differential equation and whose solutions are continuous. This was a professional break for me because I had analog computer experience and although a new hire at IBM, I was able to enjoy a certain visibility.)

Yet, there are cost-effective reasons for open-loop control systems. For example, some homes are heated by open-loop systems, requiring the homeowner to make adjustments for variations in room temperature. Purists might argue that a control system with a human in the loop is open *de facto*, but in this book we shall accept that humans can provide closed-loop control. After all, the control of a moving automobile is closed by humans and the law, in general, recognizes the adequacy of the system. Usually, operating systems are controlled by humans and their performance might well be an issue in litigation. It is then reasonable for forensic systems analysts to recognize that a human in the loop structure may be of closed form if an operator's position is responsible and accountable, and the analysts would expect such operational systems to be stable. Conversely, human negligence would indicate an open loop and possible process instability.

Despite the utility of open-loop control, the warnings of Shewhart and Chesterton, quoted at the beginning of this chapter, reinforce why we have controls to begin with. In practice, a process designed to be stable will eventually become unstable. The selection of parameters to optimize the stability of a process includes sensitivity analysis about the optimal solution in order to optimize their robustness (Arora, 1989; Taguchi et al., 1989). Wear and tear, fatigue, fragility—these are a few of the many reasons why an open-loop stability will not remain so, apart from disturbances.

The forensic systems analyst will be particularly interested in whether a control loop is closed because often the basic nature of control is verification and validation, or in simpler terms test and inspection. The latter are often regarded as non value-adding and are likely to be abandoned if the production schedule gets tight. A deliberate misinterpretation of the Lean philosophy is sometimes used to justify this adverse action. It cannot be repeated too often that a system without test and inspection is open loop.

Although the transfer function approach is invaluable in the design of a system, I said earlier that it is unlikely that discovery would contain sufficient data to reverse engineer an operating process. This would apply to its transfer function as well. However, stability data of the designed process should be available, together with associated graphs and charts. These design data could then be compared with similar operational data for an estimate of the conformity of the process performance during the period in litigation.

Why should any of this be necessary? A process is either effective or it is not. However, as discussed in Chapters 14 and 15, evidence that a process is not effective may be elusive when based solely on the existence of nonconforming product. An alternate argument is whether the process in question can ever have been effective. An unstable process cannot be effective and evidence of process stability or lack thereof is often easily determined from discovery.

11.3 Implementing Controls

The definition of an internal control can be summed up quite briefly as this: a control is a process to minimize the difference between the desired and the actual outputs of a system. This difference is called an error, and the purpose of the control is to provide a correction. The correction may reasonably be in proportion to the error itself and an arrangement of this type is therefore called *proportional control.* Alternatively or additionally, correction may be applied in proportion to the change in the error. An arrangement of this type is called *derivative control* or *integral control.*

Proportional control will always have a steady state error however, which can be reduced by increasing the control gain, but at some point an increase in gain will cause the process to oscillate or go unstable (Bower & Schultheiss, 1958; Murphy, 1965c). An effective gain can be selected and the steady state error reduced by the addition of an integrator to the control, which control is called, reasonably enough, proportional plus integral, or PI control.

Control can be likened to a golf game—you correct your short game and then your long game or your putting goes awry. Although integral control reduces the system's steady state error, it can also slow down system response time to a step input. The addition of derivative control offsets this disadvantage by improving the dynamic response of the system (Murphy, 1965d).

The three types of control together are called PID control—proportional, integral, and derivative. Figure 11.3 shows a schematic view of how PID control may be implemented in a system feedback loop. The drawing is simply an expansion of the $C(s)$ block shown earlier in Figure 11.2.

The constants, k_1, k_2, and k_3, are the variable gains in the proportional, integral, and derivative control loops. The reference signal, $r(t)$, is compared to the system output, $y(t)$, and $e(t)$ is the error or difference between them. The parameter, $d\tau$, is the sampled increment of time of the integrand value. Then $\delta(t)$ is the correction signal that feeds back into the process and is simply the sum:

$$\delta(t) = k_1 e(t) + k_2 \int e(\tau) d\tau + k_3 \frac{de(t)}{dt} \tag{11.8}$$

Figure 11.4 provides a graphic view of integral and derivative responses to a step input and suggests how the terms *lag* and *lead* came about. The step-forcing function is applied to the system at $t = 0$. A step input is a mathematical abstraction defined as an instantaneous arrival, but nothing is instantaneous in nature and I have exaggerated the slope of the step to accent the difference in

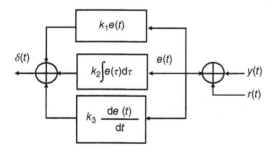

Figure 11.3 A PID feedback control structure.

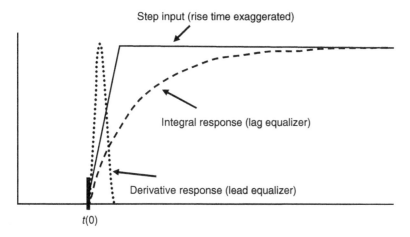

Figure 11.4 Lead and lag responses to a step input.

lead and lag responses. The derivative (lead) compensation detects and responds immediately to the step input, while the integral (lag) compensation slows down the response time. A combination of the two can be made to obtain any desired system response.

Some readers may be familiar with the terms "compensator" or "equalizer" as system controls. For example, a lag compensator may be considered as an integral control; a lead compensator may be considered as a derivative control. I have deliberately avoided using these terms because I find great confusion about them even among engineers. One group might say that a compensator will add or subtract phase margin to a closed-loop control. Another group might say that what is really needed is proper system response to an input, "proper response" being that which is viewed or recorded in real time. Thus, the argument is really over the difference between frequency and time responses, but they are simply two views of the same thing—system stability. This argument is distracting and achieves nothing. Compensator, equalizer, or controller are just alternate names for a control. Upon writing a report, the forensic systems analyst is advised to choose correct but nontechnical words where possible, and then define them at the beginning of the report. This defines the game, so to speak, and avoids quibbling over the meaning of terms.

11.4 Control of Operations

Control systems engineering is concerned with the design, development, study, and implementation of internal controls and is a field in itself with its own mathematical analyses and techniques. One might assume that such controls are primarily applied to electronic, digital, or mechanical devices such as the controls of aircraft or of production lines. This is true enough but their adaptability and application go far beyond that. During and after World War II a host of engineers began to apply systems theory and control systems design methods to social structures, among them two of my former professors, Dragoslav Šiljak (1978) and John Gibson (1977). These techniques are universal and can apply to operational structures. Organizations, too, use controls and have done so for millennia. You cannot run a railroad or an empire without controls.

The purpose of this chapter is to demonstrate the application of control theory to activities such as management, supervision, skill level, standard operating procedures, work instructions—even a signature on a sheet of paper. All follow the same rigor and formality in the design and implementation used in control engineering—all can be controls. The financial controls required by Sarbanes–Oxley are operational controls.

Operational controls follow the mathematical and physical laws of control engineering. If an operational control is not in conformance to physical laws, it probably does not work. For example, if there is no feedback in an operational structure, then there is no control in that structure. So how do the concepts of proportional, integral, and derivative control translate into operational controls?

11.4.1 Proportional (Gain) Control

As applied to organizations, proportion or gain is simply emphasis, but in its broadest sense. There is more to emphasis than pronouncements, publicity, and postings. The existence, adequacy, currency, and the consistent and proper use of procedures are manifestations of emphasis. The lack or misuse of any of these diminishes emphasis. Benign neglect and malicious compliance diminish emphasis. The forensic systems analyst is therefore interested in control gain whether in analysis of a device or of an operational activity because it affects system performance. We can say, arbitrarily, that with a properly run process that satisfies the requirements indicated in Figure 11.1, if any of the resources are missing or inadequate, we would consider the loop gain to be less than that required and the performance of the process would be suboptimal at best.

Upon reflection *authority*, too, is a control—a gain control. In Chapter 7, I quoted a requirement from Mil-Q-9858A that a person assigned to a task or responsibility should be given sufficient authority to resolve problems. Control is achieved through authority by the availability and provision of the resources required to achieve the task. This includes human, material, and capital resources, as well as time. Whether the function is performed by human or machine, "authority" aptly describes the range and power available to the function, and thus is used here in a very general sense. In systems analysis, do not think in terms of human or machine. Think in terms of function and control.

We know that authority is delegated to a person by a superior in order to achieve a task, also delegated. But the two delegations are often not in proportion because the superior may be reluctant to release authority for one reason or another. Thus, when we examine the control structure of a business organization we may see lines of responsibility in a closed form, all well and good. But the level of authority will be variable. If the person does not have sufficient authority to do the job, then the gain is less than unity.

Insufficient authority is a characteristic often found in matrix management. A project manager is given a task with specified time and budget but is rarely given the necessary physical resources to accomplish the project. Rather, this person is expected to borrow the resources from associated line managers. This works well on paper, but in reality the resources become available according to the line manager's schedule and not that of the project manager. To the forensic analyst, this is clearly a problem in control.

11.4.2 Controlling the Effect of Change

The organizational structure should be designed to detect changes in either the process input, environment, or in the process itself and to respond accordingly. In systems theory, the *environment* is that set of all events and activities that are not included in the system, but that may affect the system. Table 11.1 lists some of the events and activities within systems operations for which the risk of aperiodic or frequent change is great enough to warrant control. These events are not controls themselves, but they may require control. Some of the

Table 11.1 Some events that may require integral or rate control.

Environmental	Input	Process
Technology changes	Specifications	Common causes
Regulations	Delivery orders	Special causes
Deregulations	Change orders	Technology upgrades
Competitive innovations	Material requirements	Personnel turnover
Inflation	Material costs	Personnel absences
Demand changes	Inventory	Maintenance periods
Supply changes	Documentation requirements	Failure events
Market size	Customer complaints	Repair times
Information	Database requirements	Equipment downtime
Transport schedules	Sources	Discrepancy reports
Natural causes	Just in time activities	Training requirements

events are the reciprocal of rate—they are expressed in time periods, but the principle pertains. That is, tracking a period is equivalent to tracking a rate. Others are not rates at all, but can be converted to rates. For example, customer complaints may be converted to complaints per a given unit, say, per 1000 transactions. Another example: inventory is often rate controlled, although the rate may not be a factor of time so much as a ratio of stock on hand to a desired baseline.

Inventory rates and policies vary by industry and can be quite creative. For example, *inventory turnover rate* is a measure of how long an inventory is held in-house before it is needed. This factor, when converted to dollars, can be compared to sales in dollars and provides a measure of the cost of inventory and of the effectiveness of inventory policy. Forensic analysts usually will not have the background to evaluate this effectiveness and will be primarily concerned with whether there is an inventory policy at all. Resource management is a very specialized activity and the contribution of forensic systems analysts may be limited to advising counsel on whether a specialist might be useful in its evaluation.

There are many methods for detecting and tracking changes, among them are forecasting, market research and analysis, database management systems, information technologies, and statistical process control (SPC). These are not controls per se, but can easily be converted to controls. For example, a detection and tracking process will be developed to coordinate with a corrective process, thereby creating a control. In SPC, one such example is the Shewhart control chart, which can be used in virtually unlimited application. The forensic analyst will be concerned that rate and integral controls have been implemented and are functioning properly for any appropriate event.

11.4.2.1 Integral Control

Integration is one of the two main ideas of the calculus and refers to the sum of infinitesimal increments of a changing variable over some span of reference, usually of time, of distance, or of a dimension. As a control, integration of a process error provides a history of its change and can thereby be used to estimate and then correct for the error with the necessary increase or decrease in gain. Although integral control reduces the system's steady state error, it may also increase the rise and setting time responses to a step input because it attenuates the higher frequencies, thus slowing down system response time. This effect is indicated in Figure 11.4.

In social systems, integral control takes the form of using past history to estimate the future system state and to determine an effective correction to ensure a desired outcome. The more accurate the forecast, the smaller the steady state error. Some of the activities listed in Table 11.1 that are obvious choices for integral control would be inflation, demand, material costs, special causes, equipment down time, and parts failures.

11.4.2.2 Derivative (Rate) Control

The derivative is the other of the main ideas of the calculus and refers to the rate of change of a function of a variable with respect to the rate of change of the variable. Specifically, it examines this rate of change in the limit, as the change in the variable gets infinitesimally small. The change in the function is called a rate if the reference variable is time, although the derivative is often called a rate irrespective of the reference. For example, customer complaints per one thousand transactions may be called a rate. Since the change in rates is examined in the limit, that is, infinitesimally small, it follows that the variables must be continuous. If the variables are discrete, one refers to their rate as *differences.* Whether the rates of change are continuous or discrete is not important for this discussion; we will assume that the sampling rate for measurement will be appropriate to the rate of change being measured. A rule of thumb is that a sampling frequency should be at least 10 times faster than the sampled frequency.

In terms of cyclical operations, rate control can be obvious. However, control may be required even for aperiodic change, which may be difficult to predict, or in which errors in prediction are intolerable. In these cases, immediate detection and correction may be needed.

Derivative control is often used to complement integral control in which system response time is slowed, because it increases the system gain at higher frequencies, thus reducing system rise and settling times and improving dynamic response. This effect is indicated in Figure 11.4. Derivative control also increases the phase margin of the closed-loop system, improving system stability (Savant, 1964).

In organizational or social systems, phase margin is analogous to the lag or lead in time between an input and an output. For example, some resources have a long delivery time and must be ordered well in advance of their need. These are called, reasonably enough, "long lead time" items and the process that ensures their ready delivery is derivative control, pure and simple.

The forensic systems analyst will assume a strategic view of process control whether proportional, integral, or rate, being concerned primarily with their effectiveness and response time. As will be shown in Chapters 14 and 15, the response time of corrective action is critical to process stability, for the risk of producing nonconforming product is proportional to the period of process dysfunction.

Because PID control resolves problems with controls of conflicting capabilities, the gains of each control process must be adjusted together for optimum performance.

11.4.3 Responsibility, Authority, and Accountability

In Chapter 7 I referred to responsibility, authority, and accountability as fundamental controls, without offering any reasoning for this assignment. In fact, the three activities form a triumvirate of command. A person is assigned a task;

then is given the resources to perform the task; and then is held accountable for how well the task is achieved. In Section 11.4.1 I defined authority as the ownership of resources necessary to a task, because without the resources, authority is a meaningless word. In this sense, then, authority is a control. Therefore, responsibility and accountability are also controls.

Control is achieved through responsibility by precisely defining and assigning the task and metric of the performance function. Metric refers to the index of performance used to measure the achievement of quality in the task. Whether the function is performed by human or machine, "responsibility" aptly describes the execution and measurement of the function, and thus it is used here in a very general sense. It may sound strange to say "this machine's responsibility...," but generalizing our thoughts frees us from preconceived limitations about people and machines.

Control is achieved through accountability by imposing a penalty function for off-target performance. "Penalty" has a negative connotation here. One must account for failure to achieve or maintain the task optimally. Thus, accountability is the control used to provide the motivation to attain the goal. The penalty function is usually nonlinear, imposing a greater penalty in proportion to the deviation from target objective. Whether the function is performed by man or machine, "accountability" aptly describes the sensitivity and liability of the function, and thus it is used here in a very general sense.

Penalty function is also called *cost function*, to escape the negative connotation of the word *penalty*. It is possible to design a process with an inherent cost function. For example, the speed of an automatic lathe is controlled by a number of weighted factors; as the lathe becomes dull, the lathe will increase speed in compensation. The cost is increased by running the lathe at a higher speed. Similarly, if a project falls behind schedule, the project manager may call for the expenditure of more man hours. The cost function is the mechanism requiring the additional cost of compensation.

One might argue that the terms "responsibility" and "accountability" are redundant. This may be true in the dictionary, but it is usually not the case in industry and government. In modern organizational usage, responsibility has come to refer to the assignment of a function or task. The task may not always be achieved. For example, a production manager may be responsible for both schedule and quality control but only accountable for schedule. If the schedule slips to the point where the time for acceptance testing is impacted, the supervisor will often not hesitate to abandon the testing. Accountability means that the task *must* be achieved or a penalty is paid.

Suppose that testing is abandoned in order to avoid a late delivery. This carries with it a certain cost of quality, say customer dissatisfaction. But if a faulty part is delivered to the customer, the company will pay a much higher cost of quality than might have been the case if testing had taken place and the fault located in the shop. The customer will hold the company itself accountable for the delivery of a faulty part. Lack of accountability by the performer favors short-term penalties against long-term penalties, even though the latter almost certainly will be greater and can lead to liability. This adds a distorted and invisible element in the decision-making process.

References

ANSI/ISO/ASQ (2005). *ANSI/ISO/ASQ Q9000-2005: Quality Management Systems—Fundamentals and Vocabulary*. Milwaukee, WI: American National Standards Institute and the American Society for Quality.

Arora, J. S. (1989). *Introduction to Optimum Design*. New York: McGraw-Hill, pp. 153–167.

Bower, J. L. and Schultheiss, P. M. (1958). *Introduction to the Design of Servomechanisms*. New York: John Wiley & Sons, Inc., pp. 104–107.

Chesterton, G. K. (1908). *Orthodoxy*. New York: Lane Publishing Company.

Committee of Sponsoring Organizations of the Treadway Commission (2013). *Internal Control—Integrated Framework Executive Summary*, p. 3. https://www.coso.org/Documents/990025P-Executive-Summary-final-may20.pdf. Accessed September 19, 2017.

DeRusso, P. M., Roy, R. J., and Close, C. M. (1966). *State Variables for Engineers*. New York: John Wiley & Sons, Inc., p. 48.

Gibson, J. E. (1977). *Designing the New City*. New York: John Wiley & Sons, Inc..

Lahti, B. P. (1965a). *Signals, Systems, and Communication*. New York: John Wiley & Sons, Inc., p. 3.

Lahti (1965b), p. 14.

Luenberger, D. G. (1979). *Introduction to Dynamic Systems*. New York: John Wiley & Sons, Inc., p. 267.

Murphy, G. J. (1965a). *Control Engineering*. Cambridge, MA: Boston Technical Publishers, p. 48.

Murphy (1965b), pp. 59–60.

Murphy (1965c), p. 170.

Murphy (1965d), pp. 173–174.

Savant, C. J. (1964). *Control System Design*. New York: McGraw-Hill, pp. 218–223.

Securities and Exchange Commission (2003). *RIN 3235-A166 and 3235-A179. Management's Reports on Internal Control over Financial Reporting and Certification of Disclosure in Exchange Act Periodic Reports* (August 14, 2003). https://www.sec.gov/rules/final/33-8238.htm. Accessed September 19, 2017.

Šiljak, D. D. (1969). *Nonlinear Systems*. New York: John Wiley & Sons, Inc., p. 64.

Šiljak, D. D. (1978). *Large Scale Dynamic Systems*. New York: New Holland.

Stimson, W. A. (2005) "Sarbanes Oxley and ISO 9000." *Quality Progress*, March, pp. 24–29.

Stimson, W. A. (2012). *The Role of Sarbanes–Oxley and ISO 9000 in Corporate Management*. Jefferson, NC: McFarland, pp. 101–108.

Taguchi, G., Elsayed, A., and Hsiang, T. (1989). *Quality Engineering in Production Systems*. New York: McGraw-Hill, pp. 45–58.

United States Congress (2002). *H. R. 3763: The Sarbanes–Oxley Act of 2002*. Title IV, Section 404: Management Assessment of Internal Controls. Washington, DC: United States Congress. https://www.congress.gov/bill/107th-congress/house-bill/3763. Accessed September 19, 2017.

12

Case Study

Madelena Airframes Corporation

Forensic Systems Engineering: Evaluating Operations by Discovery,
First Edition. William A. Stimson.
© 2018 John Wiley & Sons, Inc. Published 2018 by John Wiley & Sons, Inc.

I worked as an aerospace systems engineer in industry and for the government for 25 years, 9 years in the design of control systems and 16 years in testing of large systems with embedded computers. The case study narrated in the following pages is based on real events and situations that I witnessed during this career. However, the names of persons and of corporations are fictional. The cited government agencies do exist, but their activity in this story is fictional. The story is narrated as in a novel, but it contains real manufacturing problems. All the information necessary to resolve the problems are contained in the story.

The Madelena Airframes Corporation (MAC) does not exist and never existed. It is a composite of various companies in which or with which I worked. The characteristics of the Ranger missile are a composite of various missiles, rockets, and pencil-shaped spacecraft with which I have worked. Although the roles of the persons in the story are real, the people are fictitious, as are their names that I created to suit my imagination. Any names in the story that mirror those of living persons or persons who ever lived are pure coincidence.

Nevertheless, this case study could be true. Some of the events happened, as did their consequences. The purpose of this case study is to describe the nature of evidence and procedures a forensic systems engineer or analyst might follow in the pursuit of corporate misfeasance. The format of the study is designed from the business case studies of the Darden Graduate School of Business of the University of Virginia, where I was once a student.

I am untrained in law and do not claim any expertise in that profession. In this story, however, I use a bit of law derived from my service on a jury in July 1972. The trial was presided over by Judge Melvin E. Cohn of the Superior Court of San Mateo County, CA. In this trial, there were only two witnesses: the arresting officer and the defendant, who was accused of selling drugs. The judge advised the jurors that we could not weigh more heavily the testimony of the officer over the defendant or conversely because of their positions, but if we believed that either one was lying about one thing, we could assume that he was lying about other things as well. In this fictional Madelena case, this assumption will be called the *Cohn Instruction*.

12.1 Background of the Case

MAC builds "Ranger" surface-to-air missiles for the US Navy. MAC is certified to ISO 9001 and on the basis of that certification, proposed and won an award to update its in-service missile from Version J to Version K. The latter version expands missile telemetry capability and adds a solid state test-and-operate switch. The cost of a Ranger-K is $400,000.

The Navy paid for the design, development, and fabrication of 400 Ranger-K missiles. Following fabrication, it accepted delivery of them, half of which were distributed to the fleet and the remainder evenly divided into inventories for store at the Naval Weapons Stations in Seal Beach, CA and Yorktown, VA.

In subsequent missile exercises, the first seven Ranger-Ks immediately nose-dived into the water, at which time the Chief of Naval Operations called a halt to all firings until further notice. Caught completely by surprise, the MAC top management launched an in-house investigation, meanwhile responding to Navy concerns with assurances that all delivered missiles had passed acceptance tests and suggested that the missiles may have been mishandled by Navy personnel after delivery. Following further delays and unenergetic response by MAC, the Navy called in the Department of Justice, complaining of false claims.

12.2   Problem Description

12.2.1   MAC Policies and Procedures (Missile Production)

When the design of a new missile configuration is completed, the MAC design department releases a set of drawings to both the production and test departments. Upon receiving the new drawings and having the necessary resources in inventory, Production then begins to build missiles according to these drawings. At the same time, the Test department begins to write the necessary tests of conformance to the new drawings.

Ranger missiles are built in lots to optimize equipment set-up time. Large lots receive priority over small lots, up to a feasible backlog. As 400 missiles is a very large order, the MAC production manager assigned the entire order as one lot to be fabricated. However, in keeping with its formal sampling plan of large lots, only a sample of 10% of the lot was to be tested.

12.2.2   Missile Test

Much of the test program is taken in the actual writing of the test plan, which may take up to 40 hours to write. The test engineer studies the drawings to determine all flow paths, the required stimulus devices (a variety of signal generators), the processes to be tested and their required responses, the nature of each test, and to verify circuits. The test plans are automated, but even so, the conduct of physically testing 40 missiles will also require about 80 hours to complete. As new circuitry has been added, MAC policy requires verifying product reliability by placing the missile in a chamber for a heat and vibration test, called "shake and bake." A missile has a very short life, but aerodynamic forces acting on it can cause bending and significant interior disturbance. The shake-and-bake procedure is defined in the company quality manual and usually requires about 1 hour per missile.

A missile test consists of an exercise of fin orders in yaw, pitch, and roll; verification of transmitter, receiver, and telemetry operation; and a connectivity test of the squib and of the onboard hydraulic system. The squib is a device that when submitted to an electrical charge will ignite the missile propellant.

For obvious reasons, the squib cannot be lit off in a test, so a simple resistance test is made to verify continuity from the firing circuit to the squib. This is called a connectivity test. The typical squib resistance is about 100 ohms, although this is somewhat variable.

The hydraulic system must be tested by connectivity also. The reason is that in flight, missile fins are activated by a fluid hydraulic system (FHS). At each fin command, the expended oil is spilled overboard. This feature eliminates the need for recycle equipment and, together with the diminishing oil reserve, reduces the weight of the missile in flight, thus allowing for a slightly larger warhead. For this reason, the hydraulic system cannot be used extensively to test the missile or its oil would be soon depleted, rendering the missile incapable of control.

A hydraulic test of a missile in the MAC plant is quite brief and very little oil would be consumed, but aboard ship, daily system operability tests would eventually deplete the oil reservoir. Therefore, the Ranger-K missile is designed with an internal solid state switch, called a *gate*, that serves to control missile test and operate modes. The gate offers two modes of FHS operation.

The fail safe mode is "operate." That is, the default mode of the deactivated gate is "operate" and fin orders are sent via the missile antenna to a converter/amplifier to boost up the signal, then on to the fluid hydraulic system to command the fins. However, in test mode, the gate is activated and fin orders are sent from an external hydraulic source at the test station directly to the fins. At launch, of course, all cabling to the missiles is broken, rendering the test mode irrelevant.

In addition to fin order tests, a heat and vibration test of product reliability is conducted according to company reliability policy. In the heat and vibration test, signals from new components are monitored for stability during the testing period. In regard to the test–operate gate, simply observing fin operation is insufficient; actual gate output signals must be verified.

If a failure of any kind occurs during testing, the test is stopped while a team of engineers search for the cause. In testing Rangers, there are six possibilities of failure: (i) the missile configuration is in error; (ii) one or more components are bad; (iii) one or more connections are in error; (iv) the drawings are in error; (v) the test itself is in error; and (vi) any combination of the above problems has occurred.

12.3 Examining the Evidence

The premise of this book is that there is a strong correlation between nonconforming business practices, systemic failure, product reliability, and management misfeasance. In this case study, we want to look for unusual business practices on the part of staff and management that may indicate misfeasance or malfeasance. In this evaluation, the forensic systems analyst will use an appropriate set of good business practices. For our purposes, we use ISO 9001 as a guide and want to study the evidence to answer the following questions:

- Were any procedures nonconforming? Which ones?
- Was management involved?
- Was there management negligence or misfeasance?

The evidence consists of a set of e-mails obtained by attorneys from the Department of Justice and placed in discovery. All e-mails have a Bates number for identification. The e-mails that follow are in order of Bates number and not in chronological order. For the purposes of narrating the story, they begin in the present, then switch to the past, and then switch again to the present.

12.3.1   Evidence: The Players

The evidence received by the DOJ from MAC consisted of the following e-mail. In accordance with Navy custom, the 24-hour clock is used. The number 1640 represents 4:40 p.m. The following MAC personnel played key roles in the Ranger-K problem:

1) *Victoria Falco*, General Manager of the Madalena plant in Massena, NY, builders of the Ranger family of missiles
2) *Joe Marshall*, Production manager
3) *Maria O'Brien*, Chief design engineer, Ranger-K surface-to-air missile
4) *Phil Rombaden*, Test Programs Manager
5) *Diane Maas*, Senior test engineer
6) *Alberto Salcido*, Field engineer

In addition to the MAC personnel, others involved in the problem represented the US Defense Contract Management Agency:

1) *Joan Cermak*, Madalena Resident, DCMA
2) *John Fouquier*, DCMA Headquarters

12.3.2   Evidence: E-mails

AA00123   From: Alberto Salcido
To: Victoria Falco
Cc: Joe Marshall
Subject: On-site Inspection
Date: 03 March 2015, 1640

I visited two ships at the Norfolk Naval Base that had failed Ranger-K firings. Each missile cell contained pools of oil. This is unusual, so then I visited Yorktown Weapons Station and ran continuity checks on the test circuitry of 15 Ranger-K missiles. In every missile, the wiring to the test/operate gate was inverted. In test mode, fin commands from the antenna connect to the missile hydraulic system; in operate mode, the fin commands go to the external hydraulic system, then on to the fins.

❄ ❄ ❄ ❄

AA00124   From: Phil Rombaden
            To: Victoria Falco
            Cc: Joe Marshall
            Subject: Test Schedule
            Date: 10 July 2014, 1530

Joe tells me they have 40 missiles ready for test except for the telemetry packages, which arrived this morning, two days late. They can ship the lot to Test on the 14th, but we need four or five days to write the test and about two weeks for testing. My chief test engineer has just returned from leave and has just begun writing the test. Can we get an extension?

AA00125   From: Victoria Falco
            To: Phil Rombaden
            Cc: Joe Marshall
            Subject: Test Schedule
            Date: 10 July 2014, 1830

No way, Phil. Delivery date is locked in. I'll OK the necessary overtime for your test engineer and two test teams.

AA00126   From: Phil Rombaden
            To: Diane Maas
            Subject: Test Schedule
            Date: 11 July 2014, 0930

Diane, Next Monday we'll begin to get a stream of 40 Ranger-Ks from production to test, but I can't get an extension of time to do it in. I'll authorize all the overtime you and your teams need. Sorry about that. Have you got the K-test written yet?

AA00127   From: Diane Maas
            To: Phil Rombaden
            Subject: re: test schedule
            Date: 11 July 2014, 1230

I'm working on it.

AA00128   From: Phil Rombaden
            To: Victoria Falco
            Cc: Joe Marshall
            Subject: Successful Test Program
            Date: 18 July 2014, 1930

38 of the 40 "Ks" passed their tests. In accordance with our acceptance sampling policy, the lot of 400 is GO! The two failures were sent to production for rework.

* * * * *

AA00129   From: Joan Cermak, Resident DCMA
          To: John Fouquier, DCMA Headquarters
          Subject: Ranger-K Drawings
          Date: 03 April 2015, 1445

Apparently, the Ranger-K missiles were tested with a Ranger-J test plan. No copy of a Ranger-K test plan can be found in the plant. In addition, there seems to be an error in the Ranger-K drawings. Rumor has it that wiring to the test-operate gate was inverted, but I can't get anything in writing. I've asked for a copy of the Ranger-K drawings and will check it out myself.

AA00130   From: Maria O'Brien
          To: Victoria Falco
          Cc: Phil Rombaden
          Subject: Ranger-K Retrofit
          Date: 14 April 2015, 1330

Ranger-K differed from Ranger-J by the addition of an expanded telemetry package and a solid state test/operate gate. The added circuitry required some changes in existing wiring paths to reduce the amount of new wiring required. One of the changes may have affected the gate circuitry. We tested 40 missiles. This should have been discovered in the missile testing or in shake and bake!

AA00131   From: Victoria Falco
          To: Phil Rombaden
          Subject: Test Schedule
          Date: 17 April 2015, 1545

Obviously, Ranger-K testing was botched ! I authorized 30 hours of overtime for the Ranger-K test program. How much of that time was used?

AA00132   From: Phil Rombaden
          To: Victoria Falco
          Subject: Test Schedule
          Date: 17 April 2015, 1843

Only five hours. Diane said she did not need any more and assured me that she had finished the K-test plan on the 13th of July last year and could get all the tests done in five days. I spot-checked the testing regularly, even verifying fin action on several tests. All went well, with the exception of the two failed missiles. Our acceptance plan for a sample size of 40 specifies acceptance of the lot on two failures or less; and rejection on three or more, so I passed the lot.

12.4   Depositions

In reading the depositions taken of the principal employees of MAC, the following questions may help in arriving at what really happened to cause the delivery of $160 million worth of unusable missiles. Inquiries focused on a period about mid-July 2014 when the decision to use the J-test plan was apparently made. Depositions follow, in alphabetical order of names.

12.4.1   Deposition of the General Manager

Victoria Falco: I was concerned about delays in the test program for the K missile and on 10 July 2014 talked to both Joe and Phil about the schedule. Joe told me that with overtime, production was going well and the first missiles would be sent to test on 14 July. Phil said the K-test would be ready by that time. I told him that the test program must keep on schedule and I would not accept any excuses. The accountability in the contract is quite clear: awards for staying on schedule and penalties for delays.

12.4.2   Deposition of the Senior Test Engineer

Diane Maas: I had been on leave because my child was in the hospital. I brought him home and returned to work on 10 July 2014. I met with my boss, Phil Rombaden, on the night of 11 July; it must have been about 9:30 or 10. I told him I was nowhere near ready to begin testing and couldn't possibly finish writing the K-test before three or four more days. I told him I couldn't work the week-end because my son was still recovering. Then he told me that it would be OK to use the J-test for the K-missiles. He said this might not be according to the book, but there was no fundamental problem because the test team would debug the test on the first missile, effectively creating a Version K test.

12.4.3   Deposition of the Production Manager

Joe Marshall: We were late getting the K-missiles out because of the delayed arrival of the telemetry packages. Phil and I went out to dinner on 11 July 2014 and I offered to send the first 40 missiles over to him for testing so that he could at least get the testing done. He told me he couldn't do that; the test missiles had to be drawn randomly from the entire lot. I drove him back to the plant after dinner because his car was in the parking lot. The next day Phil gave me 40 random numbers out of 400 and as soon as each of those missiles was completed, I sent it over to the test station. I heard that all but two passed their tests. As far as errors in the missiles, my people build to the drawings. If the drawings have errors, we build them in.

12.4.4 Deposition of the Chief Design Engineer

Maria O'Brien: Victoria called me in on 14 April 2015 after the fiasco with the Navy. She was pretty angry and blamed me because she thought everything started with a design error. I reminded her that's why we have test programs. I told her that per my habit, I had visited the test station late the night of 11 July 2014 to see how things were going. An acceptance test is the best in-house validation you can get of your design. No one was there but Diane; however, as I was crossing the bay, I saw Phil leave the test station and head for a side door. I guess he was going home. Diane Maas was writing the K-test and she told me everything was going fine. However, she was upset about something and finally told me her boss had just chewed her out.

12.4.5 Deposition of the Test Programs Manager

Phil Rombaden: Any of my test engineers can write a missile test, given enough time. We didn't have any time. Diane Maas is my most experienced engineer on missile check-out and has always been completely reliable. Yes, she got started on the K-test late because she had been on leave, but I knew she could get the job done on time. During the testing, I monitored the progress of her teams and it was normal. No, I did not meet with her the night of 11 July 2014. I had gone to dinner with Joe Marshall. Also, I doubt she was there. Diane didn't like working overtime because her kid was sick and she had to get home. I never gave her permission to use the J-test on the K missile and I am surprised and disappointed that this is what she apparently did.

12.5 Problem Analysis

12.5.1 Review of the Evidence

On the basis of the testimony offered by these e-mails and depositions, the forensic investigator will seek answers to these questions:

- Were any procedures nonconforming? Which ones?
- Was management involved?
- Was there management negligence or misfeasance?

The evidence suggests that Madalena's Ranger-K missile production went through acceptance testing with a Version J test plan rather than the correct Version K test plan. The "J-test" did not reveal a wiring error in the K-missile drawings and did not require a heat and vibration reliability test. As a result, the entire lot of 400 missiles was defective when delivered to the customer. Not only was the missile fin control not properly tested, but the fin control gate, improperly installed, may well have been damaged or destroyed by reverse polarity. Use of the improper test was apparently deliberate and the question arises whether this decision was made by a disgruntled employee or by management. The depositions were taken with the objective to identify the level of culpability.

Apparently, Diane Maas did not find the error in the Ranger-K drawings because she did not write a test from them. Rather than put in the very long days necessary to properly write the test, Diane simply used the test she had written for the Ranger-J, which required no "shake and bake"; no verification of test–operate signals; and no expanded telemetry test. She knew that the test team would track down any failures, so there would be no harm done. If a failure showed up during a test and was traced to the test itself, she would simply say, "Sorry guys; everyone makes mistakes." In short, she let the physical test debug the test plan rather than the other way around.

In using the older test plan while running the test, Diane did not detect that the missile hydraulic system was lit off during the test because she was watching fin action only, and very little oil was spent. She did not know about the added circuitry and did not include reliability testing and verification of stable test–operate signals. She did not know about the expanded telemetry channels and did not test for them. She tested only those channels that were common to Ranger-J and Ranger-K.

In accordance with ISO 9001, the standard invoked in the contract, top management is negligent in its failure to monitor test procedures and validate tests. However, we don't know that Diane Maas is solely responsible for failing to update the test plan. We don't know what instructions or conversations she had with her superior, except that which Rombaden admits to in his e-mail to Ms. Falco on April 17, 2015. Having her deposition from the investigation, we can compare Diane's side of the story to that of the others.

12.5.2 Nonconformities

The system audit will not necessarily be based upon a current version of a management standard, but rather its basis must be the standard version in force during the period in question. This should present no problem to the analyst, who is searching for response to test principles and not to any particular clause number or title. Once the principle is located in the relevant standard, then the analyst will cite the clause number for identification purposes during legal proceedings. As the period in litigation took place in 2014, we assume that ISO 9001 (ANSI/ISO/ASQ, 2008) is the operative standard and note the following nonconformities.

12.5.2.1 Clause 7.3.1(b) Design and Development Planning
The requirement: The organization shall plan and control the design and development of product and determine the review, verification, and validation that are appropriate to each design and development stage.

The nonconformity: Inadequate plan for management drawings review.

12.5.2.2   Clause 7.3.5 Design and Development Verification
The requirement: Product verification shall be performed in accordance with planned arrangements to ensure that the design and development outputs have met the design and development requirements.
The nonconformity: Inadequate management test design verification.

12.5.2.3   Clause 7.3.6 Design and Development Validation
The requirement: Product validation shall be performed in accordance with planned arrangements to ensure that the resulting product is capable of meeting the requirements for the specified application or intended use, where known.
The nonconformity: Inadequate management test design validation.

12.5.2.4   Clause 8.1 General Test Requirements
The requirement: The organization shall plan and implement the monitoring, measurement, analysis, and improvement processes needed to demonstrate conformity to product requirements.
The nonconformity: Inadequate management test review.

12.5.2.5   Clause 8.2.4 Monitoring and Measurement of Product
The requirement: The organization shall monitor and measure the characteristics of the product to verify that product requirements have been met.
The nonconformity: Failure to use correct procedures; failure to verify test–operate gates; and failure to conduct heat and vibration tests.

12.5.2.6   Clause 4.1 General QMS Requirements
The requirement: The organization shall determine criteria and methods needed to ensure that both the operation and control of (test) processes are effective.
The nonconformity: Failure to ensure correct test policies and procedures.

12.5.2.7   Clause 5.6.1 General Management Review Requirements
The requirement: Top management shall review the organization's quality management system, at planned intervals, to ensure its continuing suitability, adequacy, and effectiveness.
The nonconformity: Test procedure unsuitable, inadequate, and ineffective.

In my experience, top management is *always* the root cause of systemic failure because only management authority can change an errant system.

Therefore, I add Clauses 4.1 and 5.6.1 to the list because assigning noncon-formities to management clauses is an effective way to ensure that top manage-ment accepts its own responsibility and works to improve the system, rather than simply to blame subordinates. In this case, there appears to be manage-ment misfeasance. It is beyond the purview of most expert witnesses to declare negligence, but certainly they can work with attorneys to determine the feasi-bility of such charge.

12.6   Arriving at the Truth

- Who is not telling the truth?
- What is the evidence and/or the reasoning for this conclusion?
- How does this outcome affect fraud and/or false claims?

Diane Maas stated that on the night of July 11, 2014 she was ordered by her boss, Phil Rombaden, to use the J-test to test the Ranger-K missiles. The J-test failed to detect a very serious defect that rendered all 400 missiles incapable of guided flight. Any missiles that were used would be lost. Seven missiles were lost before the Navy realized the extent of the problem and stopped all future firings.

   Phil Rombaden stated that he went to dinner with the production manager on the night of July 11, 2014 and hence did not meet with Diane Maas. Rombaden claims that at no time did he order her to use the J-test for the upgraded K-missiles.

   Joe Marshall agreed that Phil and he went to dinner the night of July 11, in Joe's car, so that after dinner he returned Phil to the Madalena parking lot to retrieve his own car.

   Maria O'Brien stated that from professional habit, she had visited the test station late on the night of July 11. Although testing had not begun, she noted that Diane Maas was at work, apparently writing a test. Maria further stated that she had seen Phil Rombaden leaving the test station as she was crossing the bay. This implies he had the occasion to talk to Diane that night.

   Victoria Falco stated that she told Phil Rombaden she would tolerate no excuses for failure to finish the test program on schedule. This would put great pressure on the test manager to put proportionate pressure on Diane Maas.

   What can we make of these conflicting statements? Well, in logic, a state-ment is an expression that is either true or false. So also in forensics. There are many statements in the depositions, but two are major:

1) Phil Rombaden met with Diane Maas in the night of July 11, 2014.
2) Phil Romaden ordered Diane Maas to use the J-test on the K-missiles.

Phil Rombaden denies both (1) and (2). However, according to Joe Marshall, he had the opportunity, and according to both Maas and Maria O'Brien, he did meet with his test engineer that night, after dinner. As Rombaden lied about the first statement, then according to the Cohn Instruction, we may assume that he lied about the second. It seems that the evidence leads to malfeasance and a charge for fraud and false claims.

12.7 Damages

12.7.1 Synthesis of Damages

Four hundred Ranger-K–guided missiles were sold, paid for, and shipped to the US Navy. However, the final test of each of them was invalid because of an error in the assembly of a critical part. This would render irrelevant any subsequent portion of each test. And because of the error, seven of the missiles, fired in at-sea exercises, were uncontrollable in flight and were effectively wasted.

Relief for this complaint must take several forms. First, seven new missiles must be built, tested for connectivity and control, processed through heat and vibration tests, and then shipped to the Navy to replace those that had been lost.

Secondly, the remaining 393 missiles must complete verification, which in itself takes two forms:

1) All remaining missiles must have their test–operate control gates replaced, properly installed, and then retested. This can be done on location.
2) A sample of 40 missiles must be returned to Massena for acceptance testing of missile reliability subject to heat and vibration. While in-house, these 40 missiles can undergo gate replacement and retest, and then returned to the Navy ready for use.

The remaining 353 missiles will undergo repair and test procedures at their Navy location, including ships on deployment.

Two caveats apply here: First, if the 40-sample acceptance test fails, then all 353 missiles must be returned to Massena for heat and vibration tests according to generally accepted quality control policy (Grant & Leavenworth, 1988). However, the Navy might very well waive this QA requirement because of its cost or because of the longer delay in resolving the issue.

Secondly, the heat and vibration tests of the seven new missiles cannot be included in the 40-missile sample because they are from a different lot than the original 400. Test policy is that each lot is defined and processed individually and test integrity decorum must be maintained.

12.7.2 Costs of Correction

Several policies had to be established specifically to address correction, among them that two 2-person test and repair teams would be sent to each of six

US Navy facilities. Just one 2-person team would require overtime and too many days to complete their tasks. The six facilities are as follows: (i) Naval Station, Norfolk, VA, Atlantic Fleet Headquarters; (ii) Atlantic Fleet, deployed; (iii) Naval Station, San Diego, Pacific Fleet Headquarters; (iv) Pacific Fleet, deployed; (v) Naval Weapons Station, Yorktown, VA; and (vi) Naval Weapons Station, Seal Beach, CA.

The cost summary is shown as appropriate areas of cost, but monetary costs would depend on the market at the time. Damage strategy is determined by the attorneys in litigation and the contribution to be made by forensic systems analysts is their own best estimate of the costs of correction. This cost would be greatly increased if the criterion of the acceptance tests in heat and vibration are not met. If more than two missiles fail, then the acceptance test policy is that the entire lot must be tested. In principle, this would require the return to Massena of the 353 missiles still at issue.

- Fabrication, test, and delivery of seven Ranger-K missiles
- Shipment of 40 missiles from Navy to MAC, missile correction, test, and return
- Travel and per diem of engineers to the fleets for correction of 353 missiles
- Gate replacement and test of 393 missiles

Beyond this, the particulars of the Madelena case imply fraud and false claims, which would raise triple damages under Federal FCA laws. Or, as Walter Scott (1806) put it, "Oh, what a tangled web we weave, when first we practice to deceive."

References

ANSI/ISO/ASQ (2008). *ANSI/ISO/ASQ Q9001-2008: Quality Management Systems—Requirements*. Milwaukee, WI: American National Standards Institute and the American Society for Quality.

Grant, E. L. and Leavenworth, R. S. (1988). *Statistical Quality Control*. New York: McGraw-Hill, p. 477.

Scott, W. (1806). *Marmion: A Tale of the Flodden Field*. Edinburgh: Archibald Constable.

13

Examining Serially Dependent Processes

Many manufacturing and service processes are iterative—the units are produced or provided in a sequence one after another, although the throughput varies with the nature of the unit. It takes longer to produce an automobile than a printed circuit card and even longer to produce a ship. The output of such production systems can be said to be serial, and if the effort in each one is complete and independent of the next, so also the units should be complete and independent, one from the other.

Shewhart (1931) and others have shown that no process is perfect and that even a process in conformity to its design will suffer variation in the quality characteristics of its output because of a stable system of chance causes. However, if the process is stable, then this variation from one output to the next will be uncorrelated. Consider the variation of the quality characteristics of each produced unit as a random variable, X_j. In a stable process, there will be no correlation between, say, the value of X_j and the value of X_{j+1}. The set of these random variables is said, then, to be serially independent.

Forensic Systems Engineering: Evaluating Operations by Discovery,
First Edition. William A. Stimson.
© 2018 John Wiley & Sons, Inc. Published 2018 by John Wiley & Sons, Inc.

In an unstable process of serially produced units, there may be *autocorrelation*. The characteristic measured of each unit is often called an observation. The strength of the correlation between the units is measured by a *coefficient*, r_k, and the interval between the observations is called a lag. The autocorrelation coefficient with a lag of, say, k is

$$r_k = \frac{\sum (x_t - \bar{x})(x_{t+k} - \bar{x})}{\sum (x_t - \bar{x})^2}. \tag{13.1}$$

Equation 13.1 may serve as a definition of serial dependence. When $r_k = 0$, the units are independent—there is no correlation. When $r_k \neq 0$, there is correlation of magnitude r_k.

This definition of serial dependence has important ramifications in manufacturing processes because causality is not a factor in the characterization. Yet, correlation where there should be none always raises the question, "What is the cause?" Correlation of itself may or may not be causal. That is, correlation does not imply causation. Box et al. (1978) give a humorous example of the correlation established in the 1930s between the increase in population of a village in Germany and the increase in the number of storks observed on the rooftops. Conditional probability may suggest causation, but as in the case of storks and babies, the cause is not necessarily one of the identified correlated units. The root cause may well be what Box calls a lurking variable—not readily apparent, but an external agent acting on the correlated variables.

In statistical analysis, the assertion that correlation does *not* imply causation means that one of the correlated variables may not be the cause of a change in the other. It does not mean that there is no cause—there is always a cause of correlation where there should be none. The root cause *may* be one of the correlated variables or it may be a lurking variable. Hence, causality seems to describe two different processes. However, we shall see later in this chapter that there is only one root cause of serial dependence of product.

13.1 Serial Dependence: Causal Correlation

The correlation of produced units is often found in the creative phases of the production of systems: manufacture, assembly, and verification and validation. This is so because systems are composed of elements assembled for a synergistic purpose. Consider the case where a contractor builds cascade amplifiers. On a single printed circuit board are mounted n single pole-zero networks, each with transistor amplifier. The job consists of installing and tuning each of the n networks for minimum load to the succeeding network. Each network is tuned either correctly or not. If a network is mistuned, the subsequent network will also be mistuned, since it will be tuned to the wrong load. A mistune is a nonconformity. Hence, the probability of a nonconformity in each network is conditional upon the tuning of its predecessor as well as its own tuning.

After its n networks are tuned, the board proceeds to a test station, where the bandwidth and gain of the assembled amplifier is tested. An out-of-tolerance bandwidth or gain is a nonconformity. Thus, serially dependent workstations are of interest to forensic systems analysis. Notice that it makes no difference whether the n networks are assembled at one station, or whether one network is assembled at n stations.

This example displays the fundamental characteristic about systems mentioned earlier: a system has a synergistic objective that goes beyond the individual objectives of its elements and is therefore dependent upon them. Equally importantly, it often happens that the individual elements become dependent upon one another in meeting the requirements of synergism. It is this interdependence of elements that may show up as serial dependence in a production process organized as a workstation sequence.

The major premise of this book is that when a process is nonconforming, the key values of the units that it produces are causally correlated and serially dependent. In this chapter, the theory of serial dependence is applied to manufacturing processes in order to demonstrate this effect.

In the technical language of engineering, when a characteristic such as a dimension is measured, the result is called a *variable*. However, when a measurement is made of a characteristic such as the number of articles conforming or failing to conform to specified requirements, the result is called an *attribute* (Grant & Leavenworth, 1988). In either case, the key value of a variable or an attribute is a random variable.

In general form, serial dependence of attributes can be expressed by the following. Let an attribute, X, have a value X_t at time t, and let samples of this attribute be taken at equally spaced times, $t, t-1, t-2, ..., t-n$. Then the values of such attributes have a conditional probability expressible as

$$P(X_t \mid X_{t-1}) = f(X_t, X_{t-1}, t),$$ (13.2)

where $f(X_t, X_{t-1}, t)$ is an arbitrary function of current and previous values of the time of observation and of lag one.

Serial dependence may be costly. In the case of the cascade amplifier described earlier, if the jth network is mistuned, then the cost of quality is the cost of retuning not just the jth network but all the following networks as well. Again, consider a missile launcher that is installed on the deck of a fighting ship, then scuff-coat is applied to the deck, and finally the deck is painted. If the launcher is mispositioned by even a small amount, all three jobs must be partially undone, and then redone.

13.2 Properties of Serial Dependence

13.2.1 Work Station Definition

A work station is defined as a place and time pair in which an activity occurs. Examples of relevant activities are work, inspection, measurement, and decision. The result of an activity may be measured as an attribute of time, cost, or quality. Work sequences include (i) moving a product from station to station in an assembly line; (ii) progressing from work item to work item on the critical path of a project; or (iii) conducting a series of related tests.

Consider the n work stations depicted in Figure 13.1. The output of each work station provides the input to the following work station. Each work station output is identified as an attribute, X_j. Associated with each work station is a probability of defect, or defect rate, that is intrinsic to that station. It would exist even if the station were in isolated use. For example, let the work station be a stand-alone machine that performs tasks. The defect rate of that machine is the ratio of the number of defective tasks to the total number of tasks performed. This defect rate is denoted as the *basic defect rate, p_j*. This basic defect rate will not be known for certain; one can only estimate this parameter.

However, once the "machines" or work stations are put into the system, one cannot directly make a measurement of the p_j. When the process is in operation, one will measure at station j the correlated defect rate $\hat{p}_j$, rather than the basic, uncorrelated defect rate p_j. It will be shown that there is a way to compute p_j based upon measurements of $\hat{p}_j$. The basic, or uncorrelated, defect rate is itself an estimate, but in the interest of avoiding redundant notations, the basic defect rate is herein denoted p_j and the correlated defect rate is denoted $\hat{p}_j$.

13.2.2 Assumptions

In the following development, assumptions are made concerning the behavior of serially dependent attribute data, and from them is built a probability distribution of this behavior. From this distribution, a model is constructed that

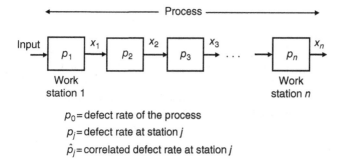

p_0 = defect rate of the process
p_j = defect rate at station j
$\hat{p}_j$ = correlated defect rate at station j

Figure 13.1 A generalized work station scenario.

describes the production process, and from the model, a process control system is developed. Thus, the foundation of this process control scheme lies in the assumptions. Serial dependence of attribute data is a special form of serial dependence, and so it must be defined. With respect to attribute data, serial dependence is defined as

$$P\left(X_j = 1 \mid X_{j-1} = 1\right) = 1; \quad \text{for all } j{:}1 < j \leq n, \tag{13.3}$$

where X_j is the attribute of the jth activity. In other words, there is a sequence of n work stations, each performing an activity, and X_j is a random variable. If a defect is found, then $X_j = 1$, else $X_j = 0$. The success of each activity depends upon the success of its predecessor. Thus, if $X_{j-1} = 0$, then X_j may or may not be in conformity, but if $X_{j-1} = 1$, then $X_j = 1$ also, because of correlation.

From this point on and in the interest of brevity, the following notation will be used: the symbol X_j implies that $X_j = 1$; and the symbol $\neg X_j$ implies that $X_j = 0$. The symbol $\neg$ implies "not," and in this context means that no defect was found at the jth activity. Starting with this description of attribute serial dependence, the necessary assumptions for determining the serial dependence of the sequence follow.

13.2.2.1 Assumption 1
Conditional dependence implies that $P(X_j \mid X_{j-1}) = 1$, where $1 < j < n$. The conditional dependence of interest is that of serial dependence.

13.2.2.2 Assumption 2
The initial defect rate p_1 is obtainable. We assume that data are available to estimate an initial basic defect rate p_1 of the first station (Station 1). If only the overall process defect rate is available from initial data, then an estimate of the basic defect rate of Station 1 may be made from this.

13.2.2.3 Assumption 3
The defect rate for the initial station is independent and is defined as the probability that $P(X_1 = 1) = p_1$. In systems theory, the environment is all that lies outside the system (Hall, 1989). Thus, being the initial element of the process, or system, the first work station receives its input from the environment. Its conformance depends only upon itself and is independent of the state, performance, or conformance of ensuing work stations of the process. Therefore, the probability of defect of X_1 is the basic, observed defect rate of Station 1, p_1.

From these three assumptions, the related probabilities are $P(X_j = 0) = q_j$, and also $P(X_j = 0 \mid X_{j-1} = 1) = 0$. One further statement is inferred from the assumptions: for a station in control, $P(X_j = 1 \mid X_{j-1} = 0) = p_j$. Use of the term "in control" does not mean that the jth work station is necessarily in control, but that it is subject to its own dynamics and is not conditioned upon the $j - 1$ work station.

13.2.3 Development of the Conditional Distribution

From the assumptions stated in the previous section, the process distribution can be determined. To develop this distribution, we will assume that each work station has the same basic defect rate, p_0, where $p_0 = P(X_j | \neg X_{j-1} = 0)$. In general this would not be true, but the assumption is made in order to determine the form of the process distribution and not to find accurate solutions of defect rates of the stations. The latter will be done later once the distribution is identified.

The process distribution is established by determining a conditional probability of defect for each station, $P(X_j | \neg X_{j-1}) = \hat{p}_j$, for all j, $1 < j < n$, where n is the number of stations. This can be done by taking advantage of the assumptions, and by constant iteration of

$$P(X_j) = P(X_j | X_{j-1})P(X_{j-1}) + P(X_j | \neg X_{j-1})P(\neg X_{j-1}). \qquad (13.4)$$

Iterating on Equation 13.4 for n stations, the conditional probability distribution expressed as $P(X_n)$, defined as $\hat{p}_n$, is

$$P(X_n) = 1 - (1 - p)^n = 1 - q_1^n \equiv \hat{p}_n. \qquad (13.5)$$

The conclusion is that the conditional structure of serially dependent attribute data increases geometrically from an initial basic defect rate, $p_0 = p_1$, at the first station. The individual $\hat{p}_j$ will reflect both serial dependence and the individual characteristics of each station. From Equation 13.5, we can compute the correlated defect rate at Station $j + 1$ due to serial dependence.

For control purposes the uncorrelated p_j is also needed, and this can be determined from statistical data according to the following argument. We will say that the measured probability is $\hat{p}_j$. Any measurement of work station defect rate necessarily measures the conditional probability of that station, since it is in the operational sequence. That is,

$$P(X_j) = \hat{p}_j = \frac{n_j}{N}, \qquad (13.6)$$

where N is the total count of the process and n_j is the number of defects at Station J over this iteration. Then starting at Station 1:

$$P(X_1) = \hat{p}_1,$$
$$P(X_2) = P(X_2 | X_1)P(X_1) + P(X_2 | \neg X_1)P(\neg X_1),$$
$$= (1)\hat{p}_1 + p_2\hat{q}_1,$$

where $P(X_2 | \neg X_1) = p_2$. As $P(X_2)$ is the conditional probability of X_2, hence $P(X_2) \equiv \hat{p}_2$. Also, $\hat{p}_2 = \hat{p}_1 + p_2\hat{q}_1$, which can be solved for the basic defect rate at Station 2:

$$p_2 = \frac{\left(\hat{p}_2 - \hat{p}_1\right)}{\hat{q}_1}.$$

Similarly, the process can be iterated over n stations with the general expression of Equation 13.6. Therefore,

$$p_j = \frac{\left(\hat{p}_j - \hat{p}_{j-1}\right)}{\hat{q}_{j-1}}. \tag{13.7}$$

Since p_j represents $P(X_j \mid \neg X_{j-1})$, then it is the uncorrelated defect rate at Station j. Finally, $\hat{p}_1 = p_1$, by Assumption 3. Thus, all the p_j can be established on a station-by-station basis.

13.2.4 Process Stability

A process subject to serial dependence is not a stationary process. The expected value for conditionally probable nonconforming units is $E[X_j] = 1 - q^j$. Consider n binary random values, X_n, each with an expected value of p_n, then

$$E[\bar{x}] = \left(\frac{1}{n}\right) \sum_{j=1}^{n} \left(1 - q^j\right) \tag{13.8}$$

For example, when Equation 13.5 is plotted with $p_0 = 0.04$, the result, shown in Figure 13.2, indicates that the probability of nonconformity of units approaches certainty, given enough work stations. Because of the increasing likelihood of nonconformity, the historical defect rate of the overall process sets a limit on the number of work stations that are feasible for that process.

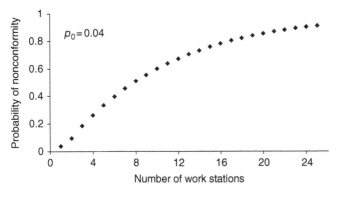

Figure 13.2 Increasing probability of nonconformity in serially dependent work stations.

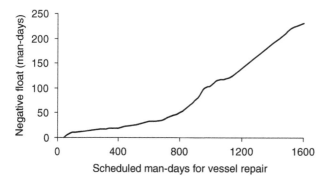

Figure 13.3 Exponential increase in negative float from serially dependent events.

This effect occurs in the presence of serial dependence regardless of the type of process and suggests the need for process control.

Consider several examples of the explosive behavior of autocorrelation in process dynamics. In a simulation study, the effect of random delays in the repair of a vessel is examined (Stimson & Mastrangelo, 1996), with the result shown in Figure 13.3. Along the critical path of the project, successive tasks are dependent on the successful completion of its predecessor. When tasks along the critical path are not completed on time, no slack (float time) is available for the activity, and the completion time of the project is subsequently delayed. This is called negative float. In Figure 13.3, the state of the repair, let us call it X_j, is shown in relation to the negative float on the critical path. Clearly, the process is unstable.

In a simulation study of a queuing process, the effect of autocorrelation exhibits the same instability (Livny et al., 1993). In this study, a rather common yet simple queuing model, called M/M/1, consists of three factors: the mean arrival rate, the mean service rate, and a single server. The customer arrival and service times are usually considered independent and identically distributed. We have all suffered in long queues and do not need to be told that if the service rate is slower than the arrival rate of customers, we may have a long wait. But with respect to serial correlation, the interesting result of this study is that the increase in wait time is exponential and increases without bound, even given moderate utilization and correlation. This destabilizing effect of correlation in serial dependence is explored graphically in Chapter 15 of this book.

Queuing theory applies to any type of line awaiting service, be it people in a bank or at a grocery store, or units in a production or assembly line. Hence, this book presents a unifying consensus of various fields of science and engineering concerning serial dependence of processes. Whether the measured random variable is the waiting time in queue, nonconformity of product, or any other subject in queue, serial dependence leads, over time, to instability of process and indeterminate or nonconforming results.

13.3   Serial Dependence: Noncausal Correlation

Serial dependence refers to whether the variables in question are serially correlated, whether causal or not. However, in engineering, there must be a cause to any effect or a problem cannot be solved. Therefore, in serial nonconformity, we conclude that the precedent is caused by the antecedent. If there is no direct cause for correlation of the variables in question, there must be somewhere a root cause, a lurking variable. In operations, the *root cause* may be the process itself, according to the following argument.

In manufacturing, a stable process will randomly produce a nonconforming unit with a probability of ≤0.0027. This is considered a rare event. Therefore, it is assumed by industrial convention that a nonconforming unit implies a special cause. Normal procedure is to stop operation and eliminate the special cause, for if not resolved, the process will be nonconforming. It will be unstable and produce a sequence of units of dubious conformity. Hence, serial dependence, the probability that a nonconforming X_j will be followed by a nonconforming X_{j+k}, is nonzero and increasing. If subsequent units from the process are tested, a correlation of random lag will be measured.

There is a subtle transition of cause here. The cause of the serial dependence of products is the nonconforming process. The cause of the nonconforming process is the special cause that, for one reason or another, was not removed. Nevertheless, it is the process that is producing the nonconforming product. It is critical to forensic analysis to distinguish the two because identifying the special cause may be illusive or even hotly denied, but the evidence of nonconformity of process cannot be denied, nor can the connection between nonconforming process and nonconforming product.

This argument applies in general. For example, in the case of the mistuned networks discussed earlier, or in the case of a defect on the critical path of a project, it is clear that a nonconforming X_j causes a nonconforming X_{j+k}. Yet, if the process is indeed stable, then X_j is the root cause only with a probability of 0.0027. Why was X_j nonconforming to begin with? Does a thing create itself? Hence, the process is always the special and root cause with a probability of 0.9973.

13.4   Forensic Systems Analysis

Evidence of serial dependence among the units of a process will usually not be found in discovery and will not be of direct concern in forensic systems analysis. The importance of serial dependence lies in its effect on the process itself—it is unstable. The purpose of this chapter is to demonstrate that a sequence of serially dependent work stations results in process instability. Unstable processes are incapable and the probability that an unstable process will produce nonconforming output increases exponentially with the duration of the instability. The root cause must be determined and eliminated.

The analyst should understand the scope of this scenario. A work station is defined as a place and time pair in which an activity occurs. Study of Figure 13.1 reveals that the set of activities constitutes the process and this is true whether one unit is processed through n work stations or whether n units are processed through one work station.

The root cause of unit nonconformity lies within the process or its environment and only rarely within the unit itself. The reason is this: once stability and capability of a process is established, then the probability that it will provide conforming output is 0.9973, by industrial convention. Thus, 99.73% of its output will be conforming. However, if the process itself becomes nonconforming, then by definition it is unstable and remains so until corrective action is taken. Its capability is no longer defined and the probability of nonconforming product increases. This will impact each unit and manifest in autocorrelation between the products as described in this chapter. Hence, the process becomes the root cause of the correlation between work stations.

The forensic analyst will examine the evidence of nonconforming processes, the period of nonconformity, and the time of initiating and completing corrective action. The conformity of all units produced during this period is in question, with the probability of unit nonconformity increasing exponentially.

Credits

This chapter is a reduction and refocus of my PhD dissertation, *Statistical Control of Serially Dependent Processes with Attribute Quality Data*, University of Virginia School of Engineering and Applied Science, Charlottesville, VA, 1994.

References

Box, G. E. P., Hunter, W. G., and Hunter, J. S. (1978). *Statistics for Experimenters.* New York: Wiley & Sons, Inc., p. 8.

Grant, E. L. and Leavenworth, R. S. (1988). *Statistical Quality Control.* New York: McGraw-Hill, p. 3.

Hall, A. D. (1989). *Metasystems Methodology.* Elmsford, NY: Pergamon Press, p. 157.

Livny, M., Melamed, B., and Tsiolis, A. (1993). "The Impact of Autocorrelation on Queuing Systems." *Management Science,* vol. 39, pp. 322–339.

Shewhart, W. A. (1931). *Economic Control of Quality of Manufactured Product.* Princeton, NJ: Van Nostrand.

Stimson, W. A. and Mastrangelo, C. M. (1996). "'Monitoring Serially Dependant Processes' with Attribute Data." *Journal of Quality Technology,* July, pp. 279–288.

14

Measuring Operations

A manufacturer or service provider offers goods and services to the public, often through a contract of performance. In the course of exercising the contract, the receiving party may determine that the performer has breached the contract in issues of time, cost, or quality of performance and product. If litigation follows and the dispute concerns the performance or management of operations, then the systems used in the operations will be challenged. But the totality of these systems is exactly the performer's quality management system (QMS). Therefore, forensic systems engineering and analysis may play a key role in the plaintiff's strategy. In this chapter, I discuss the role of the QMS in litigation and demonstrate one approach to assessment.

Discovery is a legal process used in civil lawsuits to obtain evidence not readily accessible to the applicant for use at trial, and to ascertain the existence of information that might be introduced as evidence in court. Public policy considers it desirable to give both plaintiff and defendant access to all material facts not protected by privilege to facilitate fair administration of justice. Discovery serves to level the playing field.

Tort law concerns relief for persons who have suffered injury or pecuniary damage from the wrongful acts of others. In regard to contract litigation, an argument might be made by plaintiff that the defendant failed to meet contract requirements

Forensic Systems Engineering: Evaluating Operations by Discovery,
First Edition. William A. Stimson.
© 2018 John Wiley & Sons, Inc. Published 2018 by John Wiley & Sons, Inc.

and is therefore liable for subsequent damages. In discovery, each side will attempt to gather as much information as possible, the plaintiff to show that the defendant did not meet contract requirements and the defendant to show that it did. Discovery contains the body of evidence available to both sides in the litigation.

Product liability is a special field in the practice of tort law and forensic procedures related to product failure are well established. However, there has been far less forensic activity concerning the operations and processes that produce the products. For example, in a 350 page treatise on forensic science, a committee of the National Research Council (2009) does not mention the forensics of either business operations or QMSs.

Yet, the law has been moving in the direction of recognizing business operations as entities in the law. Section 302 of the *Sarbanes–Oxley Act* (United States Congress, 2002) explicitly assigns to management the responsibility for the results of operations. Bass (2004) presented an approach to minimizing product liability through operational procedures.

Some researchers have explicitly connected quality to operations. Goodden (1996) showed how product liability could be reduced through a corporate quality system. Cook (2000) discussed the failure of complex systems, focusing on the difficulty of determining the root causes of their failure. Although implicitly tying large systems to quality, he makes no direct application to forensics. Becker and Sanborn (2001) constructed a math model to isolate the cost of a defective process on the total cost of operations. Davis (2004) showed how failure mode and effects analysis can be used to reveal "hidden factories"—those de facto processes that may not be on flow charts or that might appear benign when not. This information is useful to forensics, but the author does not expand in that direction. Boehm and Ulmer (2008) pursued the relationship of product liability to quality assurance, demonstrating that trial courts were showing increasing interest in the management of systems used in production. Berk (2009) presented procedures for failure analysis in the event of catastrophic process failure, focusing on problem solution rather than on forensic application. Russell (2013) described those areas of a corporation's supply chain most important to quality auditing, but he did not discuss the effect of process nonconformance.

A forensic systems approach to operations suggests the possibility of process liability. Attorney Leonard Miller's (1970) argument, outlined in Chapter 6 clearly establishes *process liability* in manufacturing operations. Logically, the theory of process liability is applicable to service operations also, given the definition of a system discussed in the Preface (Kalman et al., 1969). It remains to establish the role of the QMS in production and service operations.

Accepting that process liability is a concept whose time has come, what systems strategy should an attorney take in litigation? How should a forensic systems analyst approach the QMS in discovery? This chapter describes one such approach developed from experience and research into the link between systems theory and practical litigation. If the litigation concerns the effectiveness of the defendant's operations, then the defendant QMS comes into play and three questions arise:

1) What is the role of the QMS in Law?
2) Is the evidence representative of QMS performance?
3) Does forensic analysis correctly reflect the evidence?

The first question has both legal and engineering answers. The law must show that the QMS is an entity in law and an essential factor in operations performance. Engineering must show that the quality of operations and of resulting products and services depends upon the effectiveness of QMS controls.

The second question, too, has both legal and engineering answers. This question is of vital technical and statistical importance and infers technical input into the gathering of evidence. Yet, the answer is argumentative in law and requires the experience and savoir faire of attorneys, whose decisions concerning the evidence will be influenced by their litigation strategy.

The third question is technical and requires appropriate engineering analysis and synthesis. For this purpose, I consider the body of evidence as a single, if large, sample representative of the QMS and 100% of the relevant evidence is examined. Four professional fields are of particular interest in framing the evidence: research and development, operations, forensics, and law. The research activity designs a product to meet customer requirements and designs a process that can make that product with acceptable capability. Operations make the product. The forensic lab seeks to answer the question, "Was there a failure in design?" The forensic systems analyst seeks to answer the question, "Was there a failure in the process?" The attorneys in litigation seek to answer the question, "If there was a failure in product, service, or process, what is the legal basis of the failure?"

14.1   ISO 9000 as Internal Controls

We need a model to discuss the QMS and there is no loss in generality in choosing ISO 9000 (ANSI/ISO/ASQ, 2005) as that model. This set consists of three standards: ISO 9000, ISO 9001, and ISO 9004, and is recognized internationally as a set of good business practices that contribute to the quality of product and service. ISO 9001 (ANSI/ISO/ASQ, 2015) describes this activity in five categories: documentation, management, resources, operations, and verification and validation. Each requirement within these categories serves as an instruction, informing the reader on elements necessary to achieving the requirement.

I repeat from Chapter 5 that an internal control is internationally accepted as a process to assure the effectiveness of operations, reliable records and reports, and compliance with regulations. It takes but a moment's reflection to note that this definition of internal control exactly fits the requirements of ISO 9001. Therefore, ISO 9001 requirements can be regarded as internal controls for operations. Then it follows that control effectiveness can reduce the risk of liability in any operational process. One or more controls apply to each operational process. The QMS is the totality of these controls. Therefore, the QMS is the controller of operational effectiveness.

If a control conforms to QMS requirements, then it is assumed to be effective and the controlled process to be providing units in conformance to specifications (ISO 9001, Clause 8.1). If a control does not conform to requirements, then the severity of its departure must be considered. Vincins (2013) offers an index of severity with gradations suitable for litigation. *Major nonconformity* will affect unit quality or the certification of the QMS. *Moderate nonconformity* may affect unit quality or create problems in QMS certification. *Minor nonconformity* is deemed to affect neither unit quality nor company certification and would have little effect in litigation.

In this argument, the correlation of QMS effectiveness to unit quality is assumed, just as the effectiveness of good business practices or of preventive medicine is assumed. Statistical demonstrations can be made if the evidence in discovery includes a count of unit failures. In the absence of such count, a correlation is nevertheless asserted as reasonable and appropriate.

14.2 QMS Characteristics

The effectiveness of operations is often measured by variables such as flow time, throughput, utilization, and level of work in progress. These parameters might be thought of as process state variables, but they are not usually pertinent to contract litigation for two reasons. First, although they measure efficiency of process and to some extent its effectiveness, they do not measure the quality of units. Second, they reflect how things are done. A properly written standard should not tell a performer how to do something, but rather it informs on what is to be done and how achievement is to be measured (OMB, 1998).

In litigation, both plaintiff and defendant are concerned with unit quality and cost. In modern quality theory, process control is a better option than unit inspection. Effectively then, litigation is about the cost of operations and the condition of the QMS as described by management decisions, policies, and procedures. The procedures of most interest to litigants will be those of documentation, supplier control, operations, system stability, and verification and validation of product and process.

The QMS contains many controls, their effectiveness measured as attributes or as variables. This metrical mix introduces an unneeded complication in litigation that may be presented to jurors. From a contractual point of view, either a control is effective or it is not; either it is in conformity or it is not. Thus, it seems useful to measure the QMS in terms of simple attributes. Mills (1989) avers that verification is fundamentally a decision. Irrespective of whether an observation is a variable or attribute, a valid quality decision is one that is either "acceptable" or "not acceptable." This viewpoint is appropriate to forensics because it focuses on the issues in litigation, offers an easily understood metric to the jury, and because the common metric permits comparison of different population strata.

An objection may be raised that some controls should weigh more than others. Perhaps this is true and weighting can be done on a case-by-case basis, but unbiased weighting is difficult. For example, at first blush it might seem that the verification of a unit is more important than a missing signature on a document, but a nonconforming unit can be detected rather quickly and corrective action taken to limit cost. If an ineffective document is put into use, the resulting effect of nonconformity could be delayed and the cost of correction subsequently massive. As an example, in flow down a subcontractor may fail to receive a critical unit specification and, lacking this datum, provide thousands of nonconforming parts or units.

Forensics is the study of that which has already happened. Rather than weigh the relative importance of a control, a better approach is to weigh the effect of control nonconformity. The evidence will reveal whether a control is conforming or not. If conforming, the risk of defective units is considered low; if nonconforming, the risk increases. The occasions of nonconformity will be random, a pattern, or systemic. The latter is grievous and is assigned the greatest risk.

Many cases of litigation cover several years of performance, so that for statistical purposes the population of quality documentation is very large, not all of it relevant to a specific litigation. The major interest in systems forensics is in management quality decisions. There will be an effort to get the entirety of management decisions regarding policies and procedures of operations into discovery, as well as data on verification and validation. Whether or not this body of evidence approaches the totality of operational decisions depends on the litigants' response to discovery, but even if the sample is less than the total, it is usually statistically sufficient to represent those strata of the parent population.

The next question concerns the correlation of controls. Dependence is of two kinds. There is control to control, or *control dependence.* Then there is the correlation of observations of the same control taken at different times. This type of correlation is called *serial dependence,* as discussed in Chapter 13. The QMS model of this study contains five subsystems of internal controls: documentation, management, resources, operations, and verification and validation. All subsystems are designed to interact so that there is significant control dependence, which is certainly a normal occurrence in any system. Integration of diverse processes, though, invites special attention in order to avoid interface failure. But serial dependence, if it exists, is a concern.

If the QMS is stable (in control), there is little serial dependence in any of the controls (Stimson, 1994a). However, if the QMS is unstable, one or more system controls is nonconforming and there is risk of nonconforming units. The longer in time a control is nonconforming, the greater the risk of nonconforming units. The duration of system instability depends upon the response time for identifying the root cause and completing corrective action.

Worst case system instability means that nonconformity is systemic. If it is sustained, systemic nonconformity indicates management misfeasance and process liability follows. Therefore, system stability must be plausibly estimated. I define systemic nonconformity in this way: A set of nonconformities is systemic if it indicates a lack of effective process control over a period of time such that (i) unverified products may be delivered to the customer or (ii) the QMS does not meet contract requirements.

As the evidence in discovery is often gathered purposively and not randomly, a nonstatistical approach to sampling is usually suitable. Moreover, not all the evidence in discovery is taken from it; only the evidence of interest and that evidence may stand by itself, with no attempt to project it to the greater population.

14.3   The QMS Forensic Model

While there is an assumed correlation between control nonconformity and the quality of units, it cannot be estimated without information on the count of nonconforming units. However, such count is not always in evidence in discovery. It may be that only the nonconformity of controls is revealed. In this situation, forensic examination focuses on control nonconformity as a breach of contract.

The correlation between control nonconformity and unit nonconformity is estimated as a risk. Define r_{ij} as the risk assigned to the jth control in the ith subsystem. Then risk assessment can be expressed as a value between 0 and 1. Hence, when a control conforms to requirements, $r_{ij} = 0$ and there is little risk of nonconforming units. When a control is subject to special cause, then $0 < r_{ij} < 1$ and the control is subject to special cause. In this condition, there is varying risk of nonconforming units. When a control is missing or ineffective, then $r_{ij} = 1$ and nonconformity may be systemic, with high probability of nonconforming units.

Let y_i be the ith QMS subsystem. Then the risk of each subsystem is determined by a "weakest link" principle:

$$y_i = \max\left\{r_{i1}, r_{i2}, \ldots, r_{ij}, \ldots, r_{in}\right\} \tag{14.1}$$

In this model, the QMS is assigned five subsystems of differing number of controls:

y_1, risk of the documentation subsystem
y_2, risk of the management subsystem
y_3, risk of the resource subsystem
y_4, risk of the operations subsystem
y_5, risk of the verification and validation subsystem

In litigation, the subsystem risk is the inherent risk that the subsystem contributes to the provision of nonconforming units. Therefore, the risk assessment of the subsystems is similar to the risk assessment of its controls.

When there is very little risk of nonconforming units, $y_i = 0$ and the subsystem is in control. When the subsystem is in *reduced capability*, $0 < y_i < 1$, with trends or other nonrandom effects in correction mode. The risk of nonconforming units varies accordingly.

When control nonconformity is systemic with high probability of nonconforming units, then $y_i = 1$ and the subsystem is out of control.

An assessment of reduced capability depends upon the experience and background of the expert witness in analysis of quality systems. For example, if the evidence indicates that a subsystem is usually stable but suffers brief periods of nonconformity, and that there is effective corrective action, then the system could reasonably be considered in control (Steiner et al., 2005). Conversely, if the evidence indicates a subsystem has frequent periods of nonconformity varying from short to sustained and that the corrective action process is lethargic, then it could reasonably be considered out of control with high risk of liability. Reduced capability is then indicated by control risk.

14.3.1 Estimating Control Risk

As demonstrated in Stimson (1994b), the risk factors, r_{ij}, of a control are proportional to the lag in serial correlation. The greater the lag correlation within a control, the greater the risk. However, measuring this correlation presents several problems. The first is that given the probable lag epoch provided in discovery, the sample size of a given control may not meet the minimum required for reliable estimations (Box et al., 1978). The authors also warn that by the very nature of feedback in a control, serial correlation data could be unreliable.

Therefore, risk estimates based on evidence in discovery are nonstatistical, that is, they must be based on the experience and judgment of the expert witness. In this evaluation, a reliable metric for risk estimate is the response time for corrective action, which is usually available in discovery. Corrective action response time is an important requirement in any quality audit. Agreed upon by auditor and performer, it is included in an audit final report and is a benchmark for audit closure. Hence, *corrective action response* is a formal metric that must be effective and timely. As such it is useful as a proportional indicator of control risk.

Russell (1997) provides five steps in the correction process that can be used to estimate the risk that a nonconforming control may generate nonconforming units:

1) Document the problem
2) Investigate root causes
3) Design, implement, and verify the solution
4) Manage the process to avoid recurrence
5) Analyze and resolve the adverse effects of nonconformity

All of these steps must be completed within the corrective action response time agreed upon between auditor and performer.

Therefore, the risk estimates expressed earlier are reasonable and effective. If $r_{ij} = 0$, then the control conforms to requirements and there is little risk of nonconforming units. If $r_{ij} = 1$, then the control is missing or ineffective. Nonconformity may be systemic, with high probability of nonconforming units. For risks between these extremes, the estimate is based upon corrective progress in the steps described by Russell. For example, if only the first step is achieved one might estimate a high risk, say $r_{ij} \approx 0.9$, for nothing formative has been done. Conversely, if steps 1–3 are achieved, effective corrective action has been taken and the estimate might be rather low, say $r_{ij} \approx 0.3$.

Forensic analysis must arrive at a plausible estimate of correction response time for a given nonconformity if such is not in evidence. This is not as difficult as it might appear because of the cyclical nature of operations. There will be periodic operational reports, audits, test records, and quality plans; one or more of which will be issued following any nonconformity. This establishes a paper trail that outlines the nonconformity history. Corrective action would have been initiated and its progress tracked or not and if not, then the risk increases.

14.3.2 Cost of Liability

If a count of nonconforming units is available, then determination of the cost of liability is straightforward. This information is not always in evidence, in which case liability must be inferred. Suppose that the QMS is deemed out of control for a period, τ. Then the number of nonconforming units arriving at final test and inspection during this period may be significant but unknown.

The units were either tested and corrected or not at final inspection. If they were tested and corrected, then this cost may have been absorbed by the performer or have been passed on to the customer. If the former, the cost should be in discovery. If passed on to the customer and payment was requested by performer, then false claims and triple damages may be at issue in accordance with the *False Claims Act* (FCA) 31 U.S.C. §3729 (United States Code, 2007) or similar state law.

If they were delivered untested or uncorrected to the customer during the period, τ, then the customer has in inventory or in service an unknown volume of nonconforming units. The customer has no obligation to conduct the massive inspection and testing of this volume, which should have been done by the performer before delivery. On the contrary, the customer has the right to presume that units delivered under contract conform to customer requirements. Therefore, while verification of this volume is necessary, it should be conducted either by the customer or by the performer *at the cost of the performer.*

Thus, the cost of delivery to the customer of an unknown volume of nonconforming units is the cost of post delivery inspection and testing, plus the cost of correction of nonconforming units. Both costs will be even greater if the units were placed in service as components of larger systems, which must then be disassembled for inspection. Also, false claims and triple damages are at issue in accordance with the FCA or appropriate state law. A fictionalized account of just such a case is narrated in Chapter 12.

My estimates for the cost of bad quality may seem exorbitant, but they are in line with those of established manufacturing models. For example, in manufacturing, yield is the probability that a production assembly is non-defective. In their work cited earlier, Becker and Sanborn define "yielded cost" as the operating cost of a process or assembly to yield a given volume and is found by dividing accumulating costs by yield. A process yield is expressed as a fraction of good parts coming from that process. In the authors' math model, the yields appear as decimal fractions in the denominator and are multiplicative. As fractions multiply, the product becomes smaller. Hence, a very defective process will contribute a very high yielded cost to the assembly.

14.4 The Forensic Lab and Operations

In many respects, the forensic lab is akin to a research center. Both activities use formal and rigorous scientific methodologies in their work: the research center to determine an effective design; the forensic lab to determine if and why the design failed.

Forensic science can be described as the application of science and technology to investigate situations after the fact and to establish what occurred based on collected evidence. Although forensic science is often used in criminal cases, the nature of this chapter has to do with contract performance, mainly in regard to civil actions.

A forensic lab may investigate materials, products, structures, or components that fail or do not function as intended, causing personal injury or damage to property. Generally, the purpose of the investigation is to locate the causes of failure with a view to improve performance or life of a component, or to assist a court in determining the facts of an accident. If a material, product, structure, or component fails, then the cause of failure may be the process used in production and if so, the failure may be systemic. Therefore, the processes of production are properly within the purview of forensic investigation. This idea is particularly important when the failed unit is mass produced and sold widely.

The lay person might think of a forensic lab as a sort of mint—its word is gold. Law professor Brandon Garrett (2007) calls this tendency the *CSI effect*, after the TV series that has established, in the mind of the public, that forensic evidence is never wrong. However, Garrett determined that in about half of 200 DNA exonerations studied, false or misleading forensic evidence led to wrongful convictions. Gould et al. (2012) found forensic evidence error was 1 of 10 major causes of wrong convictions. These cases of misinterpreted forensics stand apart from outright criminal analyses, for example, the acts of forensic chemists that occurred in various Massachusetts science labs (Hernandez, 2013).

In short, a science lab is not infallible but uses defined processes as does any other dynamic organization. Some of the labs are even ISO 9001 certified. Within the context framed in this chapter, the forensic lab, as well as the research activity, can be considered a subset of corporate operations and its work subject to performance analysis. This systems view holds whether the lab is private or public, independent or integral to the company in litigation.

For example, a plaintiff or perhaps the defendant itself might submit a jet blade to a forensic lab for fractography, following which the lab may determine the presence of inclusions, rendering the product unusable. Given the nature of mass production, there is a possibility that many, many more such blades also have inclusions. The lab will then submit its report to the requesting party and the report will eventually appear in discovery.

Several things can go awry in this process. The lab report may be in error, or the report may be lost or abridged. On the other hand, the report may be correct and properly processed. Forensic auditors will examine the lab process and its report as they would any other document within the QMS. After all, the lab report is a comment on unit quality and system effectiveness that may indicate process liability. If the assessment of system effectiveness is judged by the conformance of internal controls in accordance with ISO 9001, then the latter becomes a de facto standard of lab results and forensic procedures.

14.5 Conclusions

Very often in a civil lawsuit, the litigation involves technical issues in which expert witnesses will be needed to pursue the case. If the technical issues relate to a performer's operations, then its QMS is at issue. Close cooperation must exist between attorney and expert witness in the assembly of evidence to ensure the availability of all evidence relevant to the litigation and to minimize the risk of judgmental error.

In regard to litigation involving corporate operations, there are three conclusions to this chapter. The first is that a QMS is properly an entity in the law. The second follows from the first: the success of litigation may depend upon attorneys' understanding of the role of the QMS in operations, for this role may be an important component of their legal strategy and will certainly help counsel to develop an effective discovery.

The third conclusion follows from the second and is that the technical community itself must be more assertive in its potential to forensics. The symbiotic relationship of product or service quality to operations must be much more widely understood. This can be accomplished by the technical community taking an active role in the propagation of its potential in forensics by contributing papers and presentations to law schools, graduate business schools, and to the legal and business communities. By its nature, the law will lag technical advancement. Process liability was first defined in 1970. It is time for forensic technology to help advance this idea.

Credits

This chapter is a reprint of "Measuring Operations with Forensic Systems Analysis," William A. Stimson (2014), *Journal of Forensic Science Policy and Management*, vol. 5, with permission of USJournalPermissions@taylorandfrancis.com.

References

ANSI/ISO/ASQ (2005). *ANSI/ISO/ASQ Q9000-2005: Quality Management Systems—Fundamentals and Vocabulary*. Milwaukee, WI: American National Standards Institute and the American Society for Quality.

ANSI/ISO/ASQ (2015). *ANSI/ISO/ASQ Q9001-2015: Quality Management Systems—Requirements*. Milwaukee, WI: American National Standards Institute and the American Society for Quality.

Bass, L. (2004). *Managing for Products Liability Avoidance*. Chicago, IL: CCH, Inc., p. 239.

Becker, D. V. and Sanborn, P. (2001). "Using Yielded Cost as a Metric for Modeling Manufacturing Processes," *Proceedings of DETC'01, ASME 2001 Design Engineering Technical Conference and Computers and Information in Engineering Conference*, Pittsburgh, PA, September 9–12, 2001.

Berk, J. (2009). *System Failure Analysis*. Materials Park, OH: ASM International, p. 6.

Boehm, T. and Ulmer, J. M. (2008). "Product Liability Beyond Loss Control—An Argument for Quality Assurance." *Quality Management Journal*, April, p. 11.

Box, G. E. P., Hunter, W. G., and Hunter, J. S. (1978). *Statistics for Experimenters*. New York: John Wiley & Sons, Inc., p. 496.

Cook, R. I. (2000). *How Complex Systems Fail.* Chicago, IL: Cognitive Technologies Laboratory, University of Chicago (Copyright by R. I. Cook, MD, for CtL), pp. 1–5.

Davis, J. (2004). *Process Failure Mode and Effects Analysis.* General Dynamics. http://www.asq0511.org. Accessed June 2, 2013.

Garrett, B. L. (2007). "Judging Innocence." *Columbia Law Review,* November 15, pp. 101–190.

Goodden, R. L. (1996). *Preventing and Handling Product Liability.* New York: Marcel Dekker, pp. 45–48.

Gould, J. B., Carrano, J., Leo, R., and Young, J. (2012). *Predicting Erroneous Convictions: A Social Science Approach to Miscarriages of Justice.* Washington, DC: National Institute of Justice.

Hernandez, C. (2013). *Thousands of Cases Compromised Due to Faulty Forensic Analysis.* Forensic Resources. North Carolina Office of Indigent Defense Services (March 13, 2013). https://ncforensics.wordpress.com/2013/03/04/thousands-of-cases-compromised-due-to-faulty-forensic-analysis/. Accessed September 16, 2017.

Kalman, R. E., Falb, P. L., and Arbib, M. A. (1969). *Topics in Mathematical System Theory.* New York: McGraw Hill, p. 74.

Miller, L. A. (1970). "Air Pollution Control: An Introduction to Process Liability and Other Private Actions." *New England Law Review,* vol. 5, pp. 163–172.

Mills, C. A. (1989). *The Quality Audit: A Management Evaluation Tool.* Milwaukee: ASQ Quality Press, p. 171.

National Research Council (2009). *Strengthening Forensic Science in the United States: A Path Forward.* Committee on Identifying the Needs of the Forensic Sciences Community. Washington, DC: The National Academies Press.

Office of Management and Budget (OMB) (1998). *A Guide to Best Practices for Performance Based Service Contracting.* https://obamawhitehouse.archives.gov/omb/procurement_guide_pbsc/. Accessed April 11, 2013.

Russell, J. P., ed. (1997). *The Quality Audit Handbook.* Milwaukee, WI: ASQ Quality Press, p. 130.

Russell, J. P. (2013). *"Keeping Watch: 'Why Supplier Audits Are Growing in Importance'."* Milwaukee, WI: ASQ Quality Press, pp. 60–62.

Steiner, S., Bovas, A., and MacKay, J. (2005). *Understanding Process Capability Indices.* Waterloo, ON: Department of Statistical and Actuarial Science, University of Waterloo, p. 12.

Stimson, W. A. (1994a). *Statistical Control of Serially Dependent Processes with Attribute Quality Data.* A dissertation to the University of Virginia School of Engineering and Applied Science, Charlottesville, VA, pp. 28–59.

Stimson (1994b), pp. 34, 55–57.

United States Code (2007). Title 31 §3729 et seq., *False Claims Act.* (January 3, 2007).

United States Congress (2002). *H. R. 3763: The Sarbanes Oxley Act,* Washington, DC.

Vincins, R. A. (2013). "Major vs. Minor." *The Auditor.* Chico, CA: Paton Press, pp. 12–14.

15

Stability Analysis of Dysfunctional Processes

The stability analysis of industrial processes may begin with Walter Shewhart's (1931a) *Economic Control of Quality of Manufactured Product*. Dr. Shewhart was a physicist at Bell Labs, concerned with reducing variation in the process of manufacturing to improve the quality of communication equipment made by Western Electric. Although the objective was to reduce the occurrence of nonconforming product, Shewhart was a systems thinker and approached the matter believing that the way to assure an improved product is to improve the process that makes the product.

Since the time of Shewhart, most research efforts have focused on improving the performance of functional systems. A cursory tour of the Internet offers hundreds of such papers in the areas of software, chemical processes, education, manufacturing, and business, to name a few. The Toyota Production System and Lean Six Sigma are examples of entire methodologies oriented to continual improvement of stable and capable systems. However, the focus of this book is on large systems that are unstable and incapable. The approach has broad application, for a system is a system and the mathematics of stability

Forensic Systems Engineering: Evaluating Operations by Discovery,
First Edition. William A. Stimson.
© 2018 John Wiley & Sons, Inc. Published 2018 by John Wiley & Sons, Inc.

analysis is substance-neutral. A dysfunctional library system behaves very much as a dysfunctional manufacturing system.

A dysfunctional system may not be observable, thus causing continued provision of questionable unit quality that may lead to litigation and forensic systems analysis. For example, many class-action suits derive from defective units that caused injury or damage to consumers. The suits may be the result of dysfunctional systems. Between 1974 and 1981, class-action suits of product liability in Federal courts increased at an average annual rate of 28% (Garvin, 1988). And they continue to rise—from 2001 to 2007, consumer protection class-action increased 50% in Federal courts (Gibson-Dunn, 2009).

Auto recalls are examples of operational but dysfunctional systems. In the Ford–Firestone litigation of 2001, the companies faced lawsuits of tens of millions of dollars and product recall costs in the billions—all because of defective tires (Schubert & Baird, 2012). In the Toyota litigation of 2010, the resulting costs were estimated at over $2 billion (Dawson, 2012).

Many federally prosecuted product liability cases are really process related. In the years 2002–2008, the US Department of Justice (DOJ, 2008) successfully prosecuted over 1300 liability cases. A Civil Division report in 2010 listed awards of over $2 billion in operations liability (DOJ, 2010).

A dysfunctional process may provide nonconforming product in large quantity, yet liability cannot be determined if evidence of such product is absent. The correlation of process to product can be established, but correlation does not prove causation. Yet, with strong correlation, domain knowledge may be sufficient to establish the causation. The purpose herein is to provide a picture of the large amount of established engineering research on the matter and to demonstrate the conditions that establish causation of a process to its product or service.

15.1 Special Terms

In litigation, it is critical that the meaning of a term be clear and unambiguous. Technical fields tend to use terms of art that differ from general understanding and even with each other. For example, the law, medicine, science, engineering, and philosophy may all have a different idea of what constitutes cause and effect. It is not my intention to challenge others, but in the interest of clarity, I have adapted several commonly used terms for purposes of forensic systems analysis. The rationale behind each definition is explained in the following paragraphs.

15.1.1 Dysfunction

There is no such thing as a perfect system. Every system suffers chance causes of variation in its output, but in a stable system, nonconforming output will be rare. Yet, there are special causes of variation that can drive the system unstable. Unable to achieve its intended purpose, such a system can be described as dysfunctional—one or more of its modes or subsystems are "out of control." As seen in many automobile recalls, instability may not be apparent if the process output is not verified and validated. A dysfunctional system is unstable and its capability cannot be defined.

Dysfunction leads to systemic process failure. W. Edwards Deming (1991a) estimated that 94% of system problems are the responsibility of management. Joseph Juran (1992) put the figure at 80% but either figure is unacceptable. Hence, the effectiveness and efficiency of a process is the responsibility of management. That effectiveness and efficiency can be enhanced and dysfunction avoided with the use of effective quality systems management.

15.1.2 Common and Special Causes

Walter Shewhart (1931a) identified two causes of product variation: "Chance (common) causes" of variation are inherent because no system is perfect. "Assignable (special) causes" of variation are those external to, but acting on, the system. Today, the more popular terms for these causes are *common* and *special* causes. Special causes may also be called disturbances, perturbations, or interventions. If the variation is bounded, the system is stable and is said to be "in control." Unbounded variation means that the system is unstable and is said to be "out of control."

15.1.3 Disturbances and Interventions

Two major types of exogenous forces act on a system: those that are desired and those that are not. For example, experimental design is an important method in cause-and-effect analysis and uses perturbation of the system as a part of test strategy. The test designer will apply a program of forces to a system and the resulting response will be mapped accordingly. This book, however, is focused on dysfunctional systems, many of which are caused by undesired perturbations. Henceforth, in this book a desired external force acting on the system will be called an intervention; an undesired force will be called a disturbance.

15.1.4 Cause and Effect

The relationship of cause and effect has been debated for 2000 years, at least. The issue is a philosophical one with a multitude of interpretations, among them that a cause–effect relationship is unknowable (Hume, 1748). Some have argued that there is no cause but only a sequence of events. Still others aver that all effects are causes initiated by the first cause. To engineers, such speculation may be entertaining but it is not practical. Engineering is based on the idea that doing a particular action results in a specific goal. Conversely, the root cause of a problem must be identified or it cannot be resolved.

Medicine, too, shares a vital interest in cause and effect because it is the basis of all medical progress. Given a set of symptoms, there is a probability of a specific malady as the cause; the greater the probability the more likely the cause. Conversely, given a specific malady, there are associated symptoms. The analysis is perhaps imperfect, but without it, medicine would be simply a game of chance.

For example, epidemiological studies examine associations between an exposure variable and a health outcome. In assessing the causal nature of an observed association, a set of criteria named after its developer, Bradford Hill, have long provided an aid to analysis (Dryer, 1994). Briefly, the Bradford Hill criteria of cause and effect are as follows:

- Strong correlation
- Consistency of evidence
- Sequence of events
- Plausibility
- Coherence with respect to natural history
- Repetition through experiment

The criteria do not claim that we can know absolutely that a given exposure causes a specific disease. There is no final proof of causality but only an inference based on an observed correspondence. Bradford Hill simply offered certain aspects of the association between an exposure and an outcome that are the most likely interpretation.

Engineering, too, takes a cautious approach to cause and effect. In operations, although the correlation of process to product may be easy to establish, causality cannot be substantiated from associations alone. Behind every causal conclusion there must lie some causal assumption that is not testable in observational studies (Pearl, 2009a).

The difficulty in establishing causality in operations is indicated by the acyclic graph of Figure 15.1. There may be hidden factors that act upon both process and product to induce characteristic variation. As the probability of identifying all possible hidden factors is remote, one cannot conclude that correlation between process and product ensures a causal relationship.

Yet, the basis of process liability is the proposition that the quality of a product depends upon the quality of the process that produces it. The usual approach in causal inference is to run randomized experiments, using interventions judiciously in analysis of response. Alternatively, given sufficient parametric data from delivered product, causality of process to product can be estimated with a certain probability.

The problem in systems forensics is that there may be insufficient evidence to estimate the character of nonconforming product, nor is it likely that the processes would be available for experiments by litigants. Responsibility in law

Figure 15.1 A causal diagram of process and product. *Source:* Stimson (2015). Reproduced with permission of Taylor & Francis.

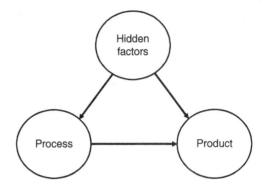

very often depends on showing that a given event or state of affairs has caused specific harm or loss to another (Honoré, 2010). If evidence is available, the forensic systems analyst can make the appropriate statistical analysis. But in the absence of evidence (nonconforming units), the harm may be unclear. The forensic problem is the more difficult where a performer uses ineffective test and inspection procedures, for the result of such testing is dubious at best. Simply put, there is no evidence of harm if no one is looking.

Although correlation does not imply causation, causation does imply correlation. Therefore, the evidence of correlation of process to product is a first and necessary condition for causality. Empirically, it must be shown that a unidirectional change in process results in a unidirectional change in product. It follows that if a change in a process can improve a product, then it may also degrade it. Theories of causation hold that causes *change the probability* of their effects; an effect may occur in the absence of a cause or fail to occur in its presence (Hitchcock, 2012). Therefore, a change followed by a trend is informative, whereas an ensuing single event may not be.

Given the constrained conditions of systems forensics, the forensic systems analyst faces several tasks: to neutralize the effects of uneven sampling found in discovery; to neutralize the presence of hidden factors in process variation; to respect accepted procedures of epidemiology; to review the various models used in engineering research on cause and effect; to demonstrate strong correlation of process and product; and to gather a sufficient body of domain knowledge as to demonstrate an overwhelming credence in engineering that the quality of a process determines the quality of the product or service that it provides.

15.2 Literature Review

The evidence gathered in discovery rarely meets the criteria of a designed experiment. Sampled evidence of process operation may be sparse and irregular in time. Aris et al. (2005) and Eckner (2014) discussed the problems of data loss and distortion because of an irregular sampling interval. These problems can be offset by offering only the most instructive evidence and by citing only those controls that are missing or ineffective for long periods. Such evidence stands by itself without recourse to the larger population.

Costa et al. (2005), Karan et al. (2006), and Wong and Lee (2010) discussed jump Markov processes with hidden disturbances or states. Such models are suitable to system forensics if they are stable, and Shewhart devised a method to neutralize the effect of hidden causes. But a dysfunctional process cannot be described by a Markov model because its transition probabilities are unknown.

Dryer (1994) contrasted the different views of the law and epidemiology relative to causality, which problem has its parallel with the law and engineering. Addressing this difference is the purpose of this chapter. Lucas and McMichael (2005) averred that the Bradford Hill criteria are guidelines rather than rules, but Höfler (2005) supported their utility, stating their heuristic value can be assessed with a counterfactual approach. As a final comment on the Hill criteria, Shewhart (1931b) satisfied them all 34 years before they were written, including the problem of hidden causes.

Illari et al. (2011) discussed two approaches to causal inference in epidemiology: (i) discovering mechanisms is necessary and sufficient for establishing a causal hypothesis and (ii) statistical analysis of associations is the primary concern. The mechanism approach supports the idea that causality can be determined with domain knowledge.

Tsay (1988) offered a model with parametric outlier, and then studied its effects on an otherwise stable system in which changes may not be obvious from the data. Reiter (2000) introduced a method to mimic randomization in observational studies by matching treatments such that they have similar distributions of causally relevant initial conditions. Feldman and Valdez-Flores (2010) discussed Markov chains with states, including operation, intervention, test, and repair that can be assigned fixed probabilities.

Faryabi et al. (2009) and Mitchell et al. (2013) utilized intervention as a state of a Markov process. Yet, if a perturbation cannot be assigned a fixed transition probability, then the process is not Markov. There is a need to distinguish between exogenous activities that can be integrated into a process as one of its states and those that are unpredictable and undesirable.

Hausman and Woodard (1999) stated that within the Causal Markov condition, genuine probabilistic dependencies reflect causal relations, which validity is inherent and necessary to the view that causes can be used to manipulate effects. Steyvers et al. (2003) showed that causal structures can be inferred from observation and intervention using a Bayesian approach with à priori probability functions relating causes to effects.

Li and Shi (2007) claimed that engineering knowledge exists from product and process designs to identify key variables and potential causal relationships. Domain knowledge plays an important role in causal discovery as it can constrain the model search, reduce the complexity, increase model accuracy, and help validate and interpret the results. In manufacturing systems, domain knowledge includes understanding of the process and product variables, their physical interactions and distributions, the production flow, procedures for data collection and quality, and the engineering specifications that support decision-making.

Pearl's (2009b) paper created a method of inference based on experimental intervention and counterfactual argument as well as on joint distributions. The idea is to exploit the invariant characteristics of structural equations without committing to a specific functional form. The invariance property permits use of a mathematical operator that simulates physical interventions by deleting certain functions from the model, replacing them with the intervention variable. The study of Vadrevu et al. (1994) showed that in regard to the wear, corrosion and other factors that contribute to degraded product quality, the time rate of change of the mean is the dominant factor. This idea, along with that of Eckner cited earlier, support my contention made later in this chapter that duration plays a crucial role in dysfunctional systems.

Yuan Qi et al. (2002) proposed a Bayesian Kalman-filter for spectrum estimation to handle unevenly sampled noisy nonstationary data. This proposal is useful to forensic analysis because the power spectrum and the autocovariance differ only by Fourier transform. Stimson (1994) showed that a serially dependent process is stable if its autocovariance generating function (AGF) converges. Conversely and pertinently, the AGF diverges if the process is unstable.

Demri et al. (2008) and Puik et al. (2013) proposed including process dysfunction in the design phase for improved reliability, an approach that resembles the *off-line design* strategy offered by Taguchi et al. (1989). This research reinforces the notion of causality between process quality and the quality of units derived from the process.

Concerned about the stability of industrial systems, Shewhart (1931c) established its theoretical and empirical manifestations. He showed that nonconformity is a by-product of an industrial process and varies according to the operational design of the process, also demonstrating that sustained serial correlation is de facto evidence of a process out of control. Miller (1970) proposed that pollution is a by-product of a manufacturing process and varies according to the operational design of the process. Box and Jenkins (1976a) used an autoregressive (AR) time series model to represent a nonstationary process out of control, causing an exponential increase in measured characteristic of output. Luenberger (1979) used a linear difference model to show how a disturbance acting upon a process can cause correlation from one output to the next and how such disturbance eventually drives a process unstable.

Deming (1991b) suggested that absent corrective action, persistent correlation between process and product can drive the process unstable. Taguchi et al. (1989) went further, explicitly showing that a process may cause characteristic variation, resulting in loss of product quality and that this loss increases monotonically from target value. They recommended periodic inspection of processes and products to ensure process stability, implying that a dysfunctional process is not always evident.

Stimson and Mastrangelo (1996) offered a theorem on stability of serially dependent processes and developed a model of serially dependent production, which was used to describe the correlation of process to product in industrial operations.

The notions of Miller and Shewhart concerning industrial by-products of pollution and variation were unified in the Chapter 14 to justify the correlation of nonconformity of process to product and to define the risks that are associated with this correlation. In this chapter, I demonstrate that these risks increase from probability to certainty if the special causes of correlation are not corrected.

15.3   Question Before the Law

In engineering circles it is generally accepted that to improve a product, one must improve the process that produces the product (Imai, 1997). A necessary corollary is that the reverse is also true: if the process degrades, the product will degrade. The unresolved question before the law is this: Does a nonconforming process necessarily provide nonconforming products? The answer seems to be yes, according to the theory of process liability, ISO 9001, and well-established engineering knowledge. Let us review these sources for this confirmation.

ISO 9001 (ANSI/ISO/ASQ, 2015) defines "product and service" to include all output categories: hardware, software, services, and processed materials. A literal reading of this definition says that these are outputs of a process. This implies that a change in the process may effect a change on the result.

The Miller theory that pollution is a by-product of the manufacturing process has an equivalent in the quality of product and service. In Chapter 9, Shewhart's seminal book on process control was summarized by Grant and Leavenworth (1988a), showing that product variation, too, is a by-product and some of it may be unacceptable. The summary is repeated here:

> Measured quality of manufactured product is always subject to a certain amount of variation as a result of chance. Some stable system of chance causes is inherent in any particular scheme of production and inspection. Variation within this stable pattern is inevitable. The reasons for variation outside this stable pattern may be discovered and corrected.

Thus, variation, too, is a by-product of the manufacturing process. When the variation is nonconforming, then Miller's argument clearly establishes *process liability* in manufacturing operations. Logically, the theory of process liability is applicable to service operations also, given the Kalman definition of a system given in the Preface (Kalman et al., 1969).

15.4   Process Stability

The correlation between the conformity of a process and the conformity of its products will affect process stability. Every product has a value or values that are critical to its intended use and to customer satisfaction. For example, a billing service might want to assure a defined minimum number of errors per 1000 billings. Or a sheet of metal might require a bolt hole at a precise location. This value is called its key characteristic (Aerospace Standard 9100, 2004), or, as abbreviated in Chapter 9, "key value." In the design phase of the product, the key values become the design values.

Following the design of the product, a process, too, must be designed that can produce the product with its design values. To do this, the process must be stable. Process stability refers to the ability of the process to make the same product repeatedly, each with its key values at design value or within acceptable limits.

As no process is perfect, this key value will vary randomly among the products due to a natural system of chance causes. But if that variation is bounded within acceptable limits, then the process is said to be stable. With a system in control, process instability should be a rare event. Recall that by industrial convention, a rare event is defined as a value in the tail areas of a normal distribution beyond $\pm 3\sigma$. Thus, the probability of a rare event is 0.0027 or less (Grant, 1988b).

Process stability is measured by taking samples of the products in small subgroups, averaging the key values of each group, then plotting the group averages as a time series. As averages are normally distributed (Grant, 1988c), the known characteristics of the normal distribution can be used to estimate the areas of stability for any process.

A typical plot of the averages of key values as a time series is called a Shewhart control chart; an example of this is shown in Chapter 9 (Figure 9.3). The process is considered stable as long as the values stay within the upper and lower control limits that indicate the boundaries of ± 3 standard deviations. The grand average (average of the averages) is also plotted to indicate trends and outliers. As long as the variation of key values stays within the control limits, no action is taken. This variation constitutes the inevitable band of chance causes described by Shewhart, including any hidden causes, and indicates that the process is stable and "in control."

There is nothing sacrosanct about ± 3 standard deviations; this industrial convention is an economic choice. Lower limits would lead to the expense of pursuing false alarms—external disturbances that do not exist; higher limits would lead to ignoring very real disturbances.

If the key values exceed the control limits, then the process is unstable. Worst case process instability occurs when the nonconformity is systemic, which I define in this way: A set of nonconformities is systemic if it indicates a lack of effective process control over a period of time such that (i) unverified products may be delivered to the customer, or (ii) the QMS does not meet contract requirements.

15.4.1 Internal Control

COSO, the Committee of Sponsoring Organizations of the Treadway Commission (2013), defines internal control as a process to assure effectiveness of operations, reliable records and reports, and compliance with regulations. It is one of the few explicit technical requirements of the Sarbanes–Oxley Act of 2002 (United States Congress, 2002). United States Securities and Exchange Commission (2003) also uses this definition, which implies its acceptance by the Federal government.

It takes but a moment's reflection to note that the definition of internal control exactly fits the requirements of ISO 9001. Therefore, ISO 9001 requirements can be regarded as internal controls for operations. Then it follows that control effectiveness can reduce the risk of liability in any operational process. One or more controls apply to each operational process. The QMS is the totality of these controls. Therefore, the QMS is the controller of operational effectiveness.

If a control conforms to QMS requirements, then it is assumed to be effective and the controlled process to be providing units in conformance to specifications (ISO 9001, ANSI/ISO/ASQ, 2015, Clause 8.1). If a control does not conform to requirements, then the severity of its departure must be considered. Vincins (2013) defines major nonconformity as that which will affect product quality or the certification of the QMS. The primary concern of the forensic lab is whether a process control is dysfunctional, for the risk of nonconforming units increases with time from probability to certainty. This is demonstrated with an autoregressive (AR) mathematical model.

15.4.2 Mathematical Model for Correlation

AR models may be used to study the serial correlation of current data to prior data of the same random variable. For example, systems of operations often perform iterative cycles called a *time series*, in which units are provided in a continuous stream. In this type of operation, samples, $x(t)$, of a unit key value are observed and recorded as a function of time, t. When used to describe stable operations, the AR model should be stationary, that is, it should vary randomly about some average value. An AR(1) model of a key value, $x(t)$, is shown in Equation 15.1. The term $(t-1)$ refers to the "lag" or interval between measurements, so if correlated, $x(t)$ is influenced by the preceding key value.

$$x(t) = \mu + \phi x(t-1) + e(t) \tag{15.1}$$

In Equation 15.1, μ represents the average value of $x(t)$ over time, ϕ is a constant proportional to the correlation between the key values of products one interval apart, and $e(t)$ is "white noise" that represents the stable system of chance causes (Box et al., 1978; Shewhart, 1931d). This equation fits a workstation scenario where observations are made of key values of products from the same process but at different times. Generally, there will be little or no correlation but only the acceptable variation of chance causes. If there is consistent correlation when there should be none, then there is a disturbance to the process that adds to chance variation. If ϕ is equal to or greater than unity, then $x(t)$ increases as a function of $|\phi|$ and will eventually be nonconforming to specifications.

Although indicating correlation, the parameter, ϕ, is not a correlation coefficient but a weight (Box & Jenkins, 1976b) and so its value can exceed unity. If $|\phi| \leq 1$, then the process is stable (Box & Jenkins, 1976c). However, if $|\phi| > 1$, then the process is unstable (*nonstationary*, in statistics) and the output quality data changes (Box & Jenkins, 1976a).

The risk that a nonconforming process will result in nonconforming units is proportional to the weight, ϕ, of the disturbance. The weight can range in value from zero to greater than unity, in agreement with control systems theory and with the Box and Jenkins model of nonstationary time series.

In *Time Series Analysis* (Box & Jenkins, 1976a), the authors use an AR(1) model to demonstrate a nonstationary process as indicative of a system out of control. Here in Figure 15.2, the Box–Jenkins result is emulated by a curve

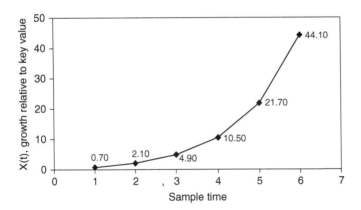

Figure 15.2 Response curve of a nonstationary AR(1) time series, $\phi = 2$. *Source:* Stimson (2015). Reproduced with permission of Taylor & Francis.

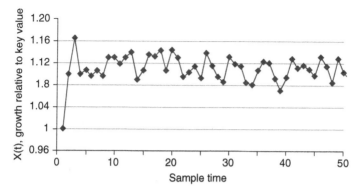

Figure 15.3 Process response to a step disturbance, $\phi = 0.1$ (weak correlation). *Source:* Stimson (2015). Reproduced with permission of Taylor & Francis.

made with a random number generator to create a stable system of chance causes. The weight is $\phi = 2$ and the curve of Figure 15.2 has the same character and similar magnitude as in the Box demonstration. In just four iterations, the curve begins a geometric increase in the product key value, indicating process instability and systemic failure. By six iterations, the series exhibits explosive behavior and completely unacceptable key values.

In Figure 15.3 we assume a weak correlation between units, say $\phi = 0.1$, caused by disturbance at time $t = 2$. The disturbance is modeled as a step function—a sudden change to the process. The system responds with a corresponding step increase in the product key value according to the magnitude of the step input. However, the process stabilizes about the new level, varying only by the white noise, $e(t)$. The process remains stable and if the step increase is small the change in the grand average may be acceptable and the system may remain uncorrected. Therefore, whether subsequent products are conforming or not depends upon the magnitude of the step relative to product specifications.

In Figure 15.4 we assume a moderate correlation, say $\phi = 0.6$. The step intervention at $t = 2$ causes the key value to increase exponentially to a new steady-state level according to the gain of the system. We now have a process that has stabilized but is possibly nonconforming because the new level is nearer to an upper control limit. The risk of the key value exceeding specification is higher. Even if the new value is deemed acceptable, the reliability of these products will be reduced because of premature wear or tolerance build-up (Taguchi et al., 1989; Vadrevu et al., 1994).

In Figure 15.5 we assume a correlation of unity—the key value of every product is dependent upon the key value of the preceding product. The current key value ramps upward to certain systemic failure, saved in this demonstration only by the corrective action taken at $t = 12$. This response at $\phi = 1$ indicates that the risk of nonconforming product is maximum.

The case in which the weight is greater than unity, $\phi > 1$, was shown earlier in Figure 15.2. The key value increases geometrically because of the exponential

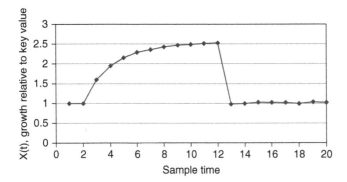

Figure 15.4 Process response to a step disturbance, $\phi=0.6$ (moderate correlation). *Source:* Stimson (2015). Reproduced with permission of Taylor & Francis.

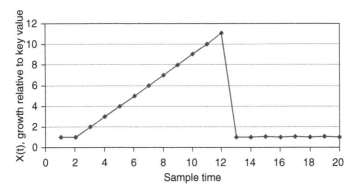

Figure 15.5 Process response to a step disturbance, $\phi=1.0$ (full correlation). *Source:* Stimson (2015). Reproduced with permission of Taylor & Francis.

increase with time of the weight itself. If a correction occurs at a later moment, then the system will return to its stable mode, but until that time the system is out of control and is providing an unknown number of nonconforming products.

15.5 Conclusions

A summary of the literature establishes two salient convictions:

1) Duration plays a crucial role in dysfunctional processes. Controls that are missing or ineffective for long periods lead to process instability and such evidence stands by itself without recourse to the larger population of evidence.
2) Engineering knowledge of product and process can establish genuine probabilistic dependencies that reflect causal relations and can identify key variables and potential causal relationships. Domain knowledge is necessary and sufficient for establishing a causal hypothesis.

Cause and effect is an essential component of human reasoning. Although causality cannot in principle be proven, given the proper conditions there is a high probability of connection. These conditions are the use of rigorous procedures in causal inference and sufficient knowledge of the domain of concern. If data are available, the forensic lab can establish the strength of correlation and the appropriate level of risk. If data are not available, the risk can be estimated with domain knowledge and understanding of serial correlation.

In this chapter, we have reviewed a wide range of research in causal inference and have discovered an overwhelming conviction within the engineering and medical communities of a causal relation between a process and its results. We have examined respected mathematical models commonly used in cause and effect studies and have determined that the causal relation is expressible in terms of risk.

Based on this study, I advance the following conclusions:

1) A process is nonconforming if one or more of its controls are nonconforming.
2) If the nonconformity is systemic, there is a risk that the process will provide nonconforming units and this risk is proportional to the delay in correcting the nonconformity. As a function of this delay, the risk will increase in probability to certainty.
3) Although often considered non-value adding, effective process verification and validation should be considered necessary activities as standard operations policy. Only in this way can process stability be ensured and dysfunction avoided.

Credits

This chapter is a reprint of "Forensic Stability Analysis of Dysfunctional Processes," William A. Stimson 2015, *Forensic Science Policy & Management: An International Journal*, vol. 6, with permission of USJournalPermissions@ taylorandfrancis.com.

References

Aerospace Standard 9100 (2004). *AS9100B, Aerospace Standard: Quality Management Systems—Aerospace—Requirements*. Washington, DC: Society of Automotive Engineer, p. 8.

ANSI/ISO/ASQ (2015). *ANSI/ISO/ASQ Q9001-2008: Quality Management System—Requirements*. Milwaukee, WI: American National Standards Institute and the American Society for Quality, p. 21.

Aris, A., Shneiderman, B., Plaisant, C., Shmueli, G., and Jank, W. (2005). "Representing Unevenly-Spaced Time Series Data for Visualization and

Interactive Exploration." *Proceedings of the 2005 IFIP TC13 International Conference on Human-Computer Interaction*, vol. 3585. College Park, MD: University of Maryland, pp. 835–846. http://hcil2.cs.umd.edu/trs/2005-01/2005-01.pdf. Accessed September 20, 2017.

Box, G. E. P. and Jenkins, G. M. (1976a). *Time Series Analysis: Forecasting and Control*. Englewood Cliffs, NJ: Prentice Hall.

Box and Jenkins (1976b), p. 51.

Box and Jenkins (1976c), p. 58.

Box, G. E. P., Hunter, W. G., and Hunter, J. S. (1978). *Statistics for Experimenters*. New York: John Wiley & Sons, Inc., p. 493.

Committee of Sponsoring Organizations of the Treadway Commission (2013). *Internal Control—Integrated Framework Executive Summary*. https://www.coso.org/Documents/990025P-Executive-Summary-final-may20.pdf. Accessed September 20, 2017.

Costa, O. L. V., Fragoso, M. D., and Marques, R. P. (2005). *Discrete-Time Markov Jump Linear Systems*. New York: Springer.

Dawson, C. (2012). "Toyota Issues Global Recall." *Wall Street Journal*, October 10. http://www.wsj.com/articles/SB10000872396390444799904578047700401681438. Accessed August 28, 2017.

Deming, W. E. (1991a). *Out of the Crisis*. Cambridge, MA: Massachusetts Institute of Technology, Center for Advanced Engineering Study, p. 315.

Deming (1991b), p. 411.

Demri, A., Charki, A., Guerin, F., and Christofol, H. (2008). "Functional and Dysfunctional Analysis of a Mechatronic System." *Reliability and Maintainability Symposium*, January 28, pp. 114–119.

Dryer, N. A. (1994). "An Epidemiologic View of Causation: How It Differs from the Legal." *International Association of Defense Counsel*. http://www.pbs.org/wgbh/pages/frontline/implants/legal/defensejournal2.html. Accessed September 20, 2017.

Eckner, A. (2014). *A Framework for the Analysis of Unevenly Spaced Time Series Data*. http://www.eckner.com/papers/unevenly_spaced_time_series_analysis.pdf. Accessed August 28, 2017.

Faryabi, B., Vahedi, G., Datta, A., Chamberland, J. F., and Dougherty, E. R. (2009). "Recent Advances in Intervention in Markovian Regulatory Networks." *Current Genomics*, vol. 10, pp. 463–477.

Feldman, R. M. and Valdez-Flores, C. (2010). *Applied Probability and Stochastic Processes*, 2nd ed. Berlin: Springer-Verlag, pp. 141–165.

Garvin, D. A. (1988). *Managing Quality: The Strategic and Competitive Edge*. New York: The Free Press, p. 23.

Gibson-Dunn (2009). *Year-End Update on Class Actions: Explosive Growth in Class Actions Continues Despite Mounting Obstacles to Certification*. Gibson, Dunn & Crutcher, LLP, February. http://www.gibsondunn.com/publications/Pages/Year-EndUpdateOnClassActions.aspx. Accessed September 17, 2017.

Grant (1988a), p. 51.

Grant (1988b), p. 60.

Grant, E. L. and Leavenworth, R. S. (1988). *Statistical Quality Control*. New York: McGraw-Hill, p. 1.

Hausman, D. M. and Woodard, J. (1999). "Independence, Invariance and the Causal Markov Condition." *British Journal of Philosophy of Science*, vol. 50, pp. 521–583.

Hitchcock, C. (2012). "Probabilistic Causation." *The Stanford Encyclopedia of Philosophy*. http://plato.stanford.edu/archives/win2012/entries/causation-probabilistic/. Accessed August 28, 2017.

Höfler, M. (2005). "The Bradford Hill Considerations on Causality: A Counterfactual Perspective." *Emerging Themes in Epidemiology*, vol. 2, p. 11.

Honoré, A. (2010). "Causation in the Law." *The Stanford Encyclopedia of Philosophy*, Winter edition. http://plato.stanford.edu/archives/win2010/entries/causation-law/. Accessed August 28, 2017.

Hume, D. (1748). *An Inquiry Concerning Human Understanding*. Great Books of the Western World, vol. 35. Chicago, IL: Encyclopaedia Britannica, Inc., pp. 451–509.

Illari, P. M., Russo, F., and Williamson, J. (2011). "Inferring Causation in Epidemiology: Mechanisms, Black Boxes, and Contrasts." *Causality in the Sciences*, edited by A. Broadbent. Oxford: Oxford University Press, pp. 45–73.

Imai, M. (1997). *Gemba Kaizen*. New York: McGraw-Hill, p. 4.

Juran, J. M. (1992). *Juran on Quality by Design*. New York: Free Press, p. 428.

Kalman, R. E., Falb, P. L., and Arbib, M. A. (1969). *Topics in Mathematical System Theory*. New York: McGraw-Hill.

Karan, M., Shi, P., and Kaya, C. Y. (2006). "Transition Probability Bounds for the Stochastic Stability Robustness of Continuous and Discrete Time Markovian Jump Linear Systems." *Automatica*, vol. 42, pp. 2159–2168.

Li, J. and Shi, J. (2007). "Knowledge Discovery from Observational Data for Process Control Using Causal Bayesian Networks." *IIE Transactions*, vol. 39, no. 6, pp. 681–690.

Lucas, R. M. and McMichael, A. J. (2005). "Association or Causation: Evaluating Links between Environment and Disease." *Public Health Classics, Bulletin of the World Health Organization*, October, pp. 792–795.

Luenberger, D. G. (1979). *Introduction to Dynamic Systems: Theory, Models, and Applications*. New York: John Wiley & Sons, Inc., p. 155.

Miller, L. A. (1970). "Air Pollution Control: An Introduction to Process Liability and Other Private Actions." *New England Law Review*, vol. 5, pp. 163–172.

Mitchell, C. M., Boyer, K. E., and Lester, J. C. (2013). "A Markov Decision Process Model of Tutorial Intervention in Task-Oriented Dialogue." *Artificial Intelligence in Education*, Lecture Notes in Computer Science, vol. 7926. Berlin: Springer, pp. 828–831. https://www.intellimedia.ncsu.edu/wp-content/uploads/mitchell-aied-2013.pdf. Accessed September 20, 2017.

Pearl, J. (2009a). "Causal Inference in Statistics: An Overview." *Statistics Surveys, Institute of Mathematical Statistics*, vol. 3, p. 99.

Pearl (2009b), pp. 96–146.

Puik, E., Telgen, D., Van Moergestel, L., and Ceglarek, D. (2013). "Structured Analysis of Reconfigurable Manufacturing Systems." *23rd International Conference on Flexible Automation and Intelligent Manufacturing* (FAIM 2013), June 26–28, 2013, Porto, Portugal.

Qi, Y., Minka, T. P., and Picara, R. W. (2002). "Bayesian Spectrum Estimation of Unevenly Sampled Nonstationary Data." *International Conference on Acoustics, Speech, and Signal Processing*, May 13–17, Orlando, FL, pp. 473–477.

Reiter, J. P. (2000). "Using Statistics to Determine Causal Relationships." *The American Mathematical Monthly*, vol. 107, pp. 24–32.

United States Congress (2002). *H. R. 3763: Sarbanes–Oxley Act of 2002.* Washington, DC: United States Congress. https://www.congress.gov/bill/107th-congress/house-bill/3763. Accessed September 20, 2017.

Schubert, D. and Baird, H. (2012). *Firestone's Tire Recall.* Daniels Fund Ethics Initiative, Anderson School of Management, University of New Mexico. Retrieved from http://danielsethics.mgt.unm.edu/pdf/Firestone%20Case.pdf. Accessed August 28, 2017.

Shewhart, W. A. (1931a). *Economic Control of Quality of Manufactured Product.* Princeton, NJ: Van Nostrand, pp. 12–14.

Shewhart (1931b), p. 130.

Shewhart (1931c), pp. 301–320.

Shewhart (1931d), p. 301.

Steyvers, M., Tenenbaum, J. B., Wagenmakers, E. J., and Blum, B. (2003). "Inferring Causal Networks from Observations and Interventions." *Cognitive Science*, vol. 27, pp. 453–489.

Stimson, W. A. (1994). "Statistical Control of Serially Dependent Processes with Attribute Quality Data." A dissertation to the University of Virginia. UMI, Ann Arbor, MI, pp. 28–59.

Stimson, W. A. (2015). "Forensic Stability Analysis of Dysfunctional Processes." *Forensic Science Policy & Management: An International Journal*, vol. 6, pp. 37–46.

Stimson, W. A. and Mastrangelo, C. (1996). "Monitoring Serially Dependant Processes with Quality Attribute Data." *Journal of Quality Technology*, July, pp. 279–288.

Taguchi, G., Elsayed, A., and Hsiang, T. (1989). *Quality Engineering in Production Systems.* New York: McGraw-Hill, pp. 12–14.

Tsay, R. S. (1988). Outliers, "Level Shifts, and Variance Changes in Time Series." *Journal of Forecasting*, vol. 7, pp. 1–20.

United States Department of Justice (2008). *Report to the President: Corporate Fraud Task Force, 2008.* Washington, DC: Office of the Deputy Attorney General, p. 5.

United States Department of Justice (2010). *Civil Division (press releases)*. https://www.justice.gov/news. Accessed September 20, 2017.

United States Securities and Exchange Commission(2003). *RIN 3235-AI66 and 3235-AI79, Management's Reports on Internal Control over Financial Reporting and Certification of Disclosure in Exchange Act Periodic Reports*. https://www.sec.gov/rules/final/33-8238.htm. Accessed September 20, 2017.

Vadrevu, S., Philipps, K., Sutherland, J., and Olson, W. (1994). "Loss Function Modeling of Time Variant Quality Characteristics." *Transactions of the North American Manufacturing Research Institution of the Society of Manufacturing Engineers*, vol. 22, pp. 343–349.

Vincins, R. A. (2013). "Major vs. Minor." *The Auditor*. Chico, CA: Paton Press, pp. 12–14.

Wong, W. C. and Lee, J. H. (2010). "Fault Detection in Process Systems Using Hidden Markov Disturbance Models." *Industrial & Engineering Chemistry Research*, 49, pp. 7901–7908.

16

Verification and Validation

I often say that when you can measure what you are speaking about and express it in numbers, you know something about it; but when you cannot measure it, when you cannot express it in numbers, your knowledge is of a meager and unsatisfactory kind: it may be the beginning of knowledge, but you have scarcely, in your thoughts, advanced to the stage of science, whatever the matter may be.

(William Thomson, Lord Kelvin, 1891)

Managers need to measure the performance of their operations for two important reasons: to evaluate the performance relative to goals and to control the processes. But Lord Kelvin goes further, suggesting that without measurement, you simply do not know what you are talking about. Although referring to science, Lord Kelvin's comments are appropriate to industrial processes. Periodic measurement

Forensic Systems Engineering: Evaluating Operations by Discovery,
First Edition. William A. Stimson.
© 2018 John Wiley & Sons, Inc. Published 2018 by John Wiley & Sons, Inc.

of industrial processes is necessary because they are dynamic and subject to change. In Chapter 13, Shewhart tells us that process variability is inevitable. In Chapter 11, Chesterton warns that a torrent of change is inevitable if a thing (process) is left to itself.

Some processes go astray more readily than others. The processes that most frequently appear in litigation of operations are design, documentation, supplier control, resource management, process management, and verification and validation (V&V, for brevity). Of these activities, the most troublesome is V&V and I have spent a considerable part of my life wondering why this should be. Intuitively, you would expect that given the cost of warranty and correction, conducting V&V prior to delivery should be automatic. And yet, in my experience this expectation is naïve. There is nothing automatic about it. On the contrary, as I shall demonstrate in this chapter, there are very strong reasons why verification and/or validation might be waived, downgraded, or simply abandoned.

16.1   Cause and Effect

Is my perception justified? Has there been a subtle but continuous downgrade of the importance of V&V in the industrial United States? If so, what are the causes and the effects to serve as evidence? To answer these questions, we need to go back to a long time ago.

16.1.1   An Historical View

In the Preface, "quality" is defined as the degree to which a set of inherent characteristics of a product or service fulfils requirements. It goes without saying that this degree must be measured, according to Lord Kelvin. Measuring the quality of products and services has been with us for centuries, since the Middle Ages at least. The famous European craft guilds of the eleventh to fifteenth centuries had as an important objective to design and maintain high standards of finished work. This quality was obtained in furniture, cathedrals, and chateaux, and we still marvel at the results.

Craftsmen were separated into three categories——master, journeyman, and apprentice——and the quality of one's work had to be demonstrated in order to gain admission to the guild. Within a shop, the quality of the product was ensured by its inspection by the master. These ideas are indicated by the Guild Act of eleventh-century England, whereby representatives of the king were invested with the power to enforce uniformity in "places of manufacture where the wardens of the crafts were appointed to see the work to be good and right and to reform what defects they should find therein, and thereupon inflict due punishment upon offenders and to stamp only good work with the seal of approval" (Hashim & Khan, 1990).

The idea of ensuring product quality by use of inspection has remained with us through the centuries and is still in use today, but inspection and tests methods have had to develop with that of the industrial age, focusing on efficiencies. David Garvin (1988) offers an interesting account of this development, particularly in American industry. Briefly, as mass production expanded, inspection of individual items became too costly. This brought about the period beginning in World War II that we call *modern quality assurance* (American Society for Quality, 2017), when reliance was placed on statistical quality control and the use of sampling.

At roughly the same time, the techniques of systems analysis were developing so that it became possible to determine methods of ensuring process quality as opposed to product quality. The basis of this change is the idea that if the process is good, the product will be good too. Industry moved from individual to random inspection of product. Quality control departments were set up independently of production, with the responsibility of ensuring final product quality through statistical means of inspections.

The transition from quality control to quality assurance (QA) was gradual, with concentration broadened to include the cost of quality, reliability engineering, and reduction of variation. But in the late 1950s, Vance Packard (1960) identified an established trend of planned obsolescence of durable goods in American industry, accompanied with a downsizing of their quality and production departments. The productivity growth that began in the postwar era stopped altogether by 1976, with the subsequent economic malaise brought about by a change in management goals and focus from production and engineering to finance and marketing (Hayes & Wheelwright, 1984a).

By 1983, the Japanese enjoyed great success in American markets with their electronic and automotive products. The key to this success was the quality of their products in contrast to the American system of planned obsolescence in which product reliability was deliberately designed to be short lived. These designs were not necessarily malevolent; a strong case can be made that Americans like new things. But liking them and being able to afford them is not the same thing and in electronics and automobiles, Americans were turning to products made in Japan. In the 1990s, American industry realized that it had to launch a counterattack of quality in response to the Japanese progress and respond it did—for a while.

This cycle of favor and disfavor of quality seems to follow the ups and downs of US industrial productivity. For example, Gavin Wright (2006) ties productivity to the gross domestic product, reporting that during the period 1950–1973, the GDP growth rate per capita averaged 2.45, while in the period 1973–1995, it fell to 1.76. However, after 1995–2000, it rose again to 2.87, about its previous level. These figures coincide with those of Figure 16.1 (Powell, 2016), which shows the rise and fall of labor productivity as a series of step functions from 1950 to 2016.

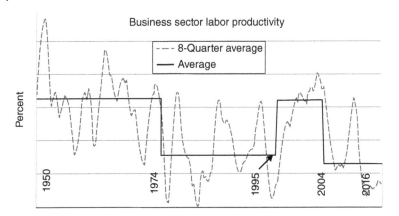

Figure 16.1 US labor sector productivity, 1950–2016.

The point to be made here is that despite the many declarations of devotion to quality and its premiere position in corporate mission statements, quality is not a sacred cow, but rather it is tied to productivity. Even then, its own rise and fall is not in proportion. Depending upon available funding, quality will lag in the rise and lead in the fall. Quality is not a peer of productivity, but it is a poor second cousin and the catalyst for second place seems almost always to be verification and validation.

16.1.2 Productivity versus Quality

In the business of making things, we would suppose that the act of making them, *production*, and how well they are made, *quality*, would go hand in hand in first place. To get to first place and stay there, metrics must be defined and measured.

Product quality is defined as the degree to which a set of inherent characteristics of a product or service fulfils requirements. The requirements are successive, beginning with the internal requirements as the product proceeds through its creative cycle, to delivery. Yet, ISO 9000 makes it clear that the ultimate requirements are those of the customer. There are several ways in which a high level of quality can be established. A company can simply declare quality through an advertising program. Many companies do this, with remarkable success. By itself this is pure deceit. The honest way to quality is to measure and improve the key characteristics of product and service through verification and validation.

Productivity is a measure of the efficiency of production. It can be defined in many ways, but the most frequently used method is to express it as a ratio of industry's output divided by the number of labor hours needed to create that

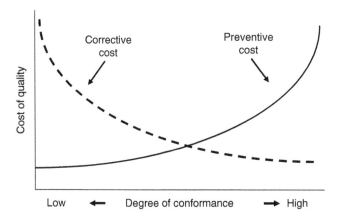

Figure 16.2 The costs of conformance.

output. Although imperfect, this definition is popular because the growth in labor productivity is a useful indicator of an economy's long-term prospects and international competitiveness (Hayes & Wheelwright, 1984b).

There are many number of books and papers in the market that claim that quality and productivity have a positive correlation; improving one can improve the other to provide happy customers and larger profits. The American Society for Quality offers over 3800 papers on this subject. Quality gurus such as Deming, Juran (1935), and Feigenbaum have written in detail on how this positive relationship can be implemented. Crosby (1979) has written that "quality is free." This title cannot be taken out of his context though, because quality certainly is not free—far from it.

Consider Figure 16.2. As the conformance of a product or process is improved, the costs of correction decrease as expected, but the costs of appraisal and prevention increase in proportion. Clearly the optimal point is at crossover, but whether the crossover is in favor of the performer, the consumer, both or neither, depends upon the static position of the performer relative to its customer base and to performance dynamics. This is invariably the difference between corporate strategy and infrastructure decisions; put another way, the difference between policy and reality.

While it is true that there *can be* a positive correlation, there can also be a negative one.

Assuming that a performer already possesses the equipment for V&V, then the only added cost to productivity is the labor necessary to conduct it. Then including V&V in the cost of labor, we can express productivity as

$$\text{Labor productivity} = \frac{\text{Value of output}}{\text{Cost of labor}}. \tag{16.1}$$

The value of output is defined by what the consumer is willing to pay. Then the value of output is the *net income* derived from the sale of products over a given period. This can be further expressed as (Stimson & Dlugopolski, 2007)

$$\text{Net income} \approx \text{Net sales} - \text{Operating expenses}, \qquad (16.2)$$

where operating expenses during the given period include the cost of verification and validation. Then truncating or abandoning V&V will decrease operating expenses, thereby increasing net income as well as the numerator of Equation 16.1. At the same time, the cost of labor will decrease, as will the denominator of Equation 16.1. Hence, the result of reducing V&V is a *doubling* of the effect on productivity and a major disincentive to pursuing quality.

When we imagine American industry, we like to think of Henry Ford and Steve Jobs, leaders who took a personal interest in their production processes and products. To quote Masaaki Imai (1997), they would often "go to Gemba," to the workplace, for hands on leadership. They were giants of quality performance.

My own favorite was Dave Packard. I worked at the Hewlett Packard facility in Palo Alto in the late 1960s, and he was frequently at the plant, asking questions, staying on top of things. And it was "Dave," not "Mr. Packard," because he understood that when you call your bosses by their first name, you are likely to tell it like it is; when you have to call your bosses by a title, you are likely to defer what you believe and say what you think they want to hear. Dave Packard understood that management cannot make good decisions with bad information.

But according to Hayes and Wheelwright (1984c), in the late 1960s, the style of management began to change. "Professional managers"—MBAs—began to fill the ranks of industry. Generally they were people with financial, accounting, marketing, and legal backgrounds rather than those with expertise in the technical aspects of manufacturing.

Productivity is a balance between production and cost. If a company's focus is on cost rather than production, then the quality of produced units is less of a consideration. Hayes asserts that short-term financial goals and marketing issues dominated the thinking of US managers in that epoch. Return on investment was the driving measure of success, focusing on the bottom line. In business in which high growth is essential, the primary task of manufacturing is "to get the product out the door."

Every company has a mission statement and a business strategy to it. That strategy must be translated to a functional strategy for operations if plan and implementation are to be unified. But it is the pattern of functional decisions actually made and the degree to which that pattern supports the business strategy that constitutes a functional strategy, and not what is said or written in annual reports or planning documents. Even in companies with a strong policy on quality, facility decisions may depart from corporate strategy when operations are in reactive mode (Hayes & Wheelwright, 1984d).

One such reactive mode is certainly a slip in schedule, especially if there are bonuses for staying on schedule. Once a performer falls behind schedule, the temptation to abandon V&V is twofold because in addition to preserving the status quo on productivity, you also improve the opportunity to get back on schedule.

Thus, there are trade-offs in operations that result in disincentives and causes that create negative correlations between quality and productivity. In regard to services, Calabrese (2013) echoes this idea, saying that a trade-off exists between service productivity and perceived service quality, an assertion supported by evidence from managerial experiences and other academic researches.

The primary champion of product and service quality, perhaps their only champion, is the customer. Yet even here there are trade-offs. Anderson et al. (1997) say that the pursuit of customer satisfaction increases costs and thereby reduces productivity. Further, they say that many economists view the relationship between productivity and customer satisfaction as generally negative. Customer satisfaction is modeled as a function of product attributes and increasing the level of the attributes increases costs and provides diminishing returns.

In litigation, the forensic systems analyst will be confronted with a vigorous assertion of corporate devotion to product quality. Plans, programs, and posters will be offered as proof that quality, like Mom's apple pie, is dear to the corporate heart and fundamental to corporate strategy. And it may well be. But at the tactical level, there are trade-offs. Verification and validation cost money and take time, often a lot of time. And the overwhelming negative cause is found in those performers who consider that V&V are non value-adding activities.

16.2 What Is in a Name?

We are in the age of word games. It is becoming common today to use a word with one intention that may be contrary to what is generally understood and may indeed be the opposite. The word *democracy* comes to mind. And so it is with measuring the quality of a product or service. Various methods are used; among them are direct run ratios, inspection, test, first article inspection, tool try, verification, validation, and more. This can be quite effective if we have a specific goal in mind and a general understanding of what is meant. However, there is room for mischief. There is room to substitute a method of lesser cost and lesser capability in lieu of the proper method. This is the arena of forensics. In current terms, the acid test lies in the meaning of verification and validation. We must keep our eye on the objective, not on the name.

When the evidence indicates that a V&V procedure was waived or abandoned, the forensic systems analyst must determine whether the customer was aware of the change and agreed to it. There is always the possibility that not having been verified, the unit under test is nonconforming. Delivery of a nonconforming unit is legal and ethical if the customer is aware of the facts and waives the relevant contract requirements. This is called acceptance by concession and many of us have done the same thing when we take our car to the garage for a checkup. We are presented with a list of defects and may very well waive correction of some of them if we regard them as minor or if we are short of cash or time.

However, when a V&V process is changed, the story is much different whether the customer knows about it or not. Invariably the change will result in the use of an inadequate procedure and the inadequacy may not be apparent to the customer, nor even perhaps to the performer. But the result is the delivery of an unverified and/or unvalidated unit. If the units are mass produced, the result may be the delivery of thousands of such units. The task of the forensic systems analyst is to determine whether the V&V process in litigation, whatever its title, was indeed capable. In the interest of staying on schedule, a shorter or faster V&V process may be substituted, which is fine if it can really do the job. But you have to get past the names of various test and inspection procedures and examine what they really do.

16.2.1 Verification and Validation Defined

During the flux and reflux of product quality, over the years there has been turmoil in exactly how it is measured, replete with a mixture of motives of how, when, and why this is done. As Machiavelli is not here to explain it to us, we must figure it out on our own. You begin with thinking about *why* things are done, and then proceed to *how* they are done. You cannot rely on the name of an activity; words can be deceptive and ambiguous. Forensic systems engineering is about the litigation of performance, which is tied to the requirements of the specifications, standards, and contracts that govern the performance. Whether a performer actually conforms to these documents often depends on how the requirements are interpreted. In Chapter 2, I discussed an attempt by standards writers to improve performance of ship repair by requiring that at least three managers would be assigned to a ship availability, assuming that more than three would be assigned to larger projects. However, this did not happen. Larger projects still received three managers because, despite the intent of the standard, three managers satisfied the words.

The analyst will rely on what, exactly, must be done to satisfy a given contracted requirement. So let us start with formal and explanatory definitions of verification and validation. We must have a reference and the most widely acceptable is the current international defining document, ISO 9000 (ANSI/ISO/ASQ, 2005). This standard defines *verification* as "confirmation through the provision of objective evidence that specified requirements have been fulfilled."

Putting a little more flesh on the bones, a product must meet customer requirements, usually expressed in specifications derived from stated expectations. Verification is a complete demonstration that the product, both in its design and in its fabrication, has the critical to quality characteristics that meet those specifications. This is achieved by design reviews, feasibility tests and demonstrations, alternative calculations, and the *appropriate measurements of the key characteristics.*

ISO 9000 defines *validation* as "confirmation through the provision of objective evidence that the requirements for a specific intended use or application have been fulfilled." At first blush this definition appears redundant, but the delineating part is the reference to *intended use.* Careful reading (which lawyers do by trade) reveals that validation is a confirmation that the specifications do indeed conform to customer requirements. Thus, validation of a unit necessarily follows its verification. This is reasonable because the performer must first demonstrate that a unit can be made before its use can be validated. The unit remains undefined until it meets the specifications.

Thus, validation has a dual nature. The first is to ensure that the verification process itself is correct according to established policies and procedures. This includes defined criteria for review and approval of the processes; approval of equipment and qualification of personnel; the use of specific methods and procedures; and maintenance of records. The second is the act of demonstrating that the product or service meets the customer's intended use requirements. Validation therefore follows verification because it "verifies" the verification process itself as well as validating the unit that has been verified. The order verification > validation is immutable.

The double duty of validation is distinguished in ISO 9001 by different clauses devoted to process validation and product validation. I make no attempt to quote clause numbers because the ISO numbering system is like shifting sand. Each version of ISO 9001 has a different numbering from the previous one. It is incumbent on forensic systems analysts to be willing to pursue information by diligent reading of documents. Both sides of litigation will do this, one in complaint; the other in defense. Very often in law, word management is the name of the game.

16.2.2 Inspection and Test

The acts of verifying and validating product and service quality have long been conducted by corporate functions called "inspection and test." Recognizing the universally understood terminology, the first international standard of quality management, ISO 9001:1987 (ANSI/ASQC, 1987), included a major requirement entitled, "Inspection and Testing." ISO 9000:2005 defined *inspection* as "conformity evaluation by observation and judgment accompanied as appropriate by measurement, testing, or gauging." The same standard defines *test* as "determination of one or more characteristics according to a procedure."

16.2.3 Monitor and Measure

The definitions of inspection and test are less precise in objective than those of verification and validation and in the year 2000 version, they disappeared, to be replaced by "monitor and measure." The term *monitor* remains undefined, but *measurement process* is defined in ISO 9000 as a "set of operations to determine the value of a quantity."

The lack of a definition for *monitor* is regrettable because its absence permits inadequate observations. I once witnessed a facility in which one person was assigned to oversee 13 operating machines that continually manufactured sheets of product. During the trek from machine to machine, no product was monitored; while observing at 1 machine, 12 others were producing unmonitored product. Considering the production rate of the machines and the flaw rate per yard of sheet, a few calculations revealed that the likelihood of observing a flaw before delivery was very small. Yet, the contractor could and did justify a claim of monitoring the process.

It is possible to construct a definition of monitor from the standard, which says on page 2, "Monitoring the satisfaction of interested parties requires the evaluation of information relating to the perception of interested parties as to the extent to which their needs and expectations have been met." This can be rewritten. Thus, *monitor* is defined as "Monitoring an object or process requires an evaluation of the extent to which needs and expectations have been met." Although not elegant, this definition would suffice to show that the observer of 13 machines was not monitoring them adequately. However, I have no idea whether this definition would stand up in court or even if an attorney would be willing to take on the challenge.

Then in the year 2008 version, although the terms monitor and measure are retained, a new set of terms were introduced: "verification and validation." Why this sequence of changes in terminology? After all, there is only one way to verify and validate a thing and that is by inspection or testing. So why the change in the names? It would seem that calling actions by one name last year and yet another this year and still another next year simply generates confusion on what, exactly, is to be done.

Presumably, the change was meant to be cosmetic. The terms "inspection and testing" are reminders of a bygone era of the early 1900s when Frederick Taylor introduced his time and motions studies. This period suffers negative connotations today and whether the terms were abandoned because of this negative sense or because new terms tend to emphasize a new era in quality is not a key issue. The key issue is whether the acts of inspecting and testing, under any name, have diminished in importance?

16.2.4 Subtle Transitions

Why is it that "V&V" is so troublesome? I have alluded to the answer in several chapters, commenting on modern quality fashions such as Six Sigma and Lean and on the general perception that test and inspection are non value-adding. However, Lean is but the latest in a trend of the postmodern period that accords with several wise sayings that somehow have been subverted. The first of these is "You cannot inspect quality into a product" (Deming, 1991). The second is "Quality is everyone's job" (Pyzdek & Keller, 2003). The third is "Inspection and testing are non value adding activities" (Liker, 2004).

Considered separately, each of these ideas is true. The first case points out the obvious: inspection comes after the fact; the unit is already made. If it is defective, inspecting will not make it better. Therefore, the focus of attention should be on making the unit defect free in the first place. However, the unintentional result was to lessen the importance of the inspection and test function to those who wanted to do so and instead place greater reliance on defect free production.

However, shifting the focus from inspection and test to defect free production simply transfers the costs of inspection to increased costs in production, which is contrary to the primary objective of improving productivity. Taking advantage of the *nihil obstat* of the quality authorities that it is acceptable to reduce test and inspection activities, many operations managers simply eliminated these functions and reduced their QA department to a one person manager with no staff.

The second, attributed to the quality guru Armand Feigenbaum, also makes sense, but it is taken out of context. Feigenbaum correctly understood that when something is everyone's responsibility, it likely never gets done. His original statement was to establish a *system of accountability* such that quality is everyone's job. Given a general aversion to accountability, the first part of this idea was simply truncated. QA was further diminished because now everybody is responsible for their own work. Why do we need QA? Under this interpretation, persons are now accountable who cannot possibly be held so in a court of law, and objectivity is demanded where objectivity is rare.

The third sage advice, "inspection and testing are non value adding activities," is the most damaging of all, as I have recounted in several chapters. In Chapter 3, I pointed out that the Toyota Production System, which is the basis of Lean, assigns a set of tasks in three categories: value added, non value added, and non value added but necessary. Validation and verification fall into the third category.

In the United States, very often Lean has only two categories: value added and non value-added, with V&V assigned to the second set. Then if the production schedule slips, V&V are truncated or abandoned under the guise that the performer, being very modern, is also very Lean. I am reminded of the good old days of the Middle Ages in France, when the king announced himself, *Très Chrétien* (Very Christian). How can you challenge such magnificence?

In Chapter 11, I pointed out that if V&V are abandoned, then you have an open-loop process and such a process is not in control. Although often considered non value-adding, the contrary is true. Effective process verification and validation should be considered necessary activities as standard operations policy. Only in this way can process stability and capability be ensured and dysfunction avoided.

How is it possible that wise sayings produce negative outcomes? Well, when you are on the production line, the schedule is slipping and the customer is screaming, a different kind of wisdom takes over. In the trenches, corporate strategy is not the rule. Survival is the rule. Or to paraphrase James Madison (1788), if people were angels, no verification or validation would be necessary. If angels were to govern people, no internal controls on operations would be necessary.

16.3   The Forensic View of Measurement

One of the challenges in forensic systems analysis of operations is to determine whether the product or service delivered to the customer met all requirements. The acid test lies in the meaning of measurement, which is the essence of verification and validation.

16.3.1   Machine Tools and Tooling

Measurement is the essence of verification and devices used to impart dimensions and other measureable characteristics must themselves be measured against some standard. In mass production, machine tools fall into this category and will often become the object of forensic review in litigation of operations. Within the context of this discussion, the purpose of *tooling* is best described as "Tooling is, in its simplest sense, the means of production. Special tooling, such as dies and molds, is custom designed and made to manufacture specific products, generally in quantity, and to desired levels of uniformity, accuracy, quality, and interchangeability" (National Tooling and Machining Association, 2011).

The need for machine tools and tooling derives from the development of mass production: they enable the production of goods in large quantities at low cost per unit. Mass produced goods can have acceptable quality by standardizing the means of precision manufactured, interchangeable parts.

The material basis for mass production was laid by the development of the machine tool industry—the making of machines to make machines. Machine tools assure precision shaping, dimensioning, and the production of interchangeable parts in large quantity. But first, the tool or jig must be correctly made; therefore, every tool is, in its creation, a product and must itself undergo verification and validation. The device does not become a tool and cannot be used as a tool until it is verified and validated.

16.3.2 Measurement

Verification and validation require that a measurement be taken. Lamprecht (1993) offers a definition of *measurement* that is suitable to our purpose:

> The quantitative determination of a physical magnitude by comparison with a fixed magnitude adopted as the standard, or by means of a calibrated instrument. The result of measurement is thus a numerical value expressing the ratio between the magnitude under examination and a standard magnitude regarded as a unit.

Although seemingly very technical, this definition is quite general. For example, it applies to measurements of uncertainty. Key characteristics may vary randomly, so that statistical techniques are required in the measurement. Despite the uncertainty, a quantitative determination of the mean and variance of the quality of the unit being measured can be made. Shewhart charts of product variation are one well-known technique of measuring process stability. The definition applies also to measurements of subjective performance. Customer evaluations, such as *good, fast, important*, and so on, can be perfectly valid measures of performance even though they are qualitative. Qualitative evaluations can be measurements if some sort of numerical scale is assigned to them.

In the provision of product or service, there are basically two types of measurements: those of the product or service (units) and those of the processes that provide them. The two are tied together in such way that one supports the other. There is an axiom in industrial statistics that it is possible for a good process to provide a bad unit, and it is possible for a bad process to provide a good unit. In statistics, a sample size of one proves nothing.

The parameters to be measured of the product or service are its key values. The parameters to be measured of the process are its stability and capability. They are tied together in this sense: first it must be demonstrated by measuring its key values that the unit can be made at all. Then a number of units must be verified to demonstrate the stability and capability of the process, which demonstration enables an estimate of the rate of good units thereafter. Subsequently, as long as the process remains unchanged and stable, further unit verification is unnecessary, although process stability and capability should be verified periodically.

There are general considerations that ensure measurement integrity and if overlooked, then the measurements taken are invalid. The essential components of measurement are objectivity, observability, parameters to be measured, methodology, metrics, validity, parametric range, sensitivity, statistical analysis, instrumentation, test specifications, and standards. All must be appropriate to the specific measurement. Some of the most frequently used methodologies in manufacturing follow.

16.3.3 Control Charting

I repeat from Chapter 9 that process stability is measured by taking random samples of the key values, measuring each of them, then plotting the values as a time series. Sampling is done in small subgroups of 2–5 units each and the average value of the subgroup is plotted. As long as the variation of a key value stays within control limits, no action is taken. This variation constitutes a band of chance causes that are inevitable and acceptable. The chart defines the bounds of the variation and indicates that the process is stable. The process is said to be "in control."

How many samples are required? Because of the way samples are taken in control charting, the distribution of unit variation is approximately normal and about 30 samples suffice, each of a handful of unit key values (McClave & Dietrich, 1988). Hence, the axiom referred to earlier about chance production of good and bad units is reinforced because it takes a number of individual measurements of units to establish the stability of the process.

Following this initial trial, process capability can be determined as described in Chapter 9. If the capability is acceptable, then process stability and capability are established with the same experiment and the system is ready for operation. Shewhart's control charting is a fundamental measurement of operations and must be remeasured whenever the process is no longer in control or has been changed. Forensic analysts should be particularly aware of the latter because sometimes small, temporary changes may seem technically or economically insignificant, but if they change process dynamics, you have a different, unverified process.

16.3.4 First Pass Yield

First Pass Yield (FPY) is determined from the initial yields of a series of processes without rework, hence the name "first pass." At each process a ratio is measured: the number of good units less the number of defective units, divided by the total number entering the process over a specific time. This decimal fraction is a measure of the good output of that process. Then the series of decimal fractions are multiplied to give the FPY, the overall rating. Suppose that there are k processes in the system. Then the FPY is

$$\text{FPY} = \prod_{i=1}^{k} \frac{x_i - d_i}{x_{i-1}} \tag{16.3}$$

where x_0 is the initial input population, x_i is the number of good units from process i, and d_i is the number of defective units at process i. Calculated in this way, FPY measures the effectiveness of the overall system by taking the product of the yields of each process in the system. As the multiplicands are decimal fractions, their product will be less than any individual process yield, hence a system measurement.

You can get the same result by simply dividing the total number of good units at the end of the sequence by the total number at its start. However, Equation 16.3 parses the system effectiveness into subsystems effectiveness, which inquiry may be useful to forensic analysis.

Notice that FPY is a process verification and does not, itself, verify or validate the units, which necessarily must be completed before one can classify the units as good or defective. Granted that good and bad units are counted, but nothing is said about *how* the units are determined to be good or bad. Once a sufficiency of units is verified and FPY is acceptable, the test is akin to a control chart for the period in question. FPY does not explicitly measure stability or capability, but given a sufficient experience and knowledge of the system, management may determine from a change in the yield if there is a destabilizing problem.

16.3.5   First Article Inspection

SAE Aerospace Standard AS9102 (SAE Aerospace, SAE International Group, 2004) defines First Article Inspection (FAI) as "A complete, independent, and documented physical and functional inspection process to verify that prescribed production methods have produced an acceptable item as specified by engineering drawings, planning, purchase order, engineering specifications, and/or other applicable design documents." The standard also states that the purpose of the first article inspection is to provide objective evidence that all engineering specification and design requirements are correctly understood, accounted for, verified, and recorded. AS9102 may be used to verify conformance of a prototype part to design requirements.

FAI requirements are common in the aerospace industry and in Federal government contracting, and they are becoming increasingly popular in general industry. FAI is a very detailed comparison of product design versus production result to ensure that the first product of a lot produced is 100% correct before continuing the production run. In sum, a full FAI is a validation of the production process and a verification of the first unit.

FAI can become a concern in forensic systems review first because of its complexity, secondly because there are legitimate questions of what it can and cannot do. It is not a test of process verification, establishing neither process stability nor capability. Also, a process must be verified and validated before its units can be so. The verification of a single unit says nothing about the process or its ability to repeat the event.

Because the procedure is very comprehensive, the standard allows the use of partial FAI procedures, which can raise another forensic issue. A partial FAI can be perfectly appropriate, but it can also be abused. For example, a pseudo FAI would be a visual inspection of the first unit of a lot, the action to be called "FAI." Performers must justify the omission of any part of a full and formal FAI.

I concede that a full FAI strongly suggests a stable and capable system of production. Nevertheless, every production system is stochastic and FAI will verify only one unit. The axiom applies here too and you cannot establish a probability distribution with a sample of one. A forensic systems review would include determining whether subsequent verification and validation of additional units were done and process stability and capability established.

16.3.6 Tool Try

The V&V of machine tools are particularly essential because of the multiplier effect of tooling. A nonconforming tool can result in dozens or hundreds of nonconforming parts. Recall from Chapter 9 that a tool remains a product until it is verified and validated. The product-to-be tool is verified by visual inspection, measurement of its dimensions and form, and by testing it in operation. The operational test is sometimes called Tool Try, which test also validates the tool. Some companies might piggyback Tool Try with a verification of the product using FAI. This is invalid. The tool/product must be verified and validated before it can be used as a tool.

There may be an industry standard for machine tool verification that the forensic systems analyst can use as a V&V reference in a particular litigation. If not, the analyst has little alternative but to determine exactly what the tool must be able to measure, then determine whether the tool in question was properly verified and validated. Moreover, the proper order is essential: tool verify, then validate; first product verify and validate; process stability and capability verify and validate. You cannot rely on the name of the test, which may vary from one performer to another.

References

American Society for Quality (2017). *History of Quality*. Milwaukee, WI: American Society for Quality. http://asq.org/learn-about-quality/history-of-quality/overview/overview.html. Accessed September 17, 2017.

Anderson, E. W., Claes, F., and Rust, R. T. (1997). "Customer Satisfaction, Productivity, and Profitability: Differences between Goods and Services." *Marketing Science*, vol. 16, no. 2, p. 131.

ANSI/ASQC (1987). *ANSI/ASQC Q91-1987. American National Standard: Quality Systems—Model for Quality Assurance on Design/Development, Production, Installation and Servicing.* Milwaukee, WI: American Society for Quality Control.

ANSI/ISO/ASQ (2005). *ANSI/ISO/ASQ Q9000-2005: Quality Management Systems—Fundamentals and Vocabulary.* Milwaukee, WI: American National Standards Institute and the American Society for Quality.

Calabrese, A. (2013). "Quality versus Productivity in Service Production Systems: An Organizational Analysis." *International Journal of Production Research*, vol. 51, no. 22, pp. 6594–6606. Published online: July 15, 2013.

Crosby, P. (1979). *Quality Is Free.* New York: McGraw-Hill.

Deming, W. E. (1991). *Out of the Crisis.* Cambridge, MA: Massachusetts Institute of Technology, p. 29.

Garvin, D. (1988). *Managing Quality.* New York: The Free Press, pp. 3–20.

Hashim, M. and Khan, M. (1990). "Quality Standards: Past, Present, and Future." *Quality Progress*, June, pp. 56–59.

Hayes, R. H. and Wheelwright, S. C. (1984a). *Restoring Our Competitive Edge: Competing through Manufacturing.* New York: John Wiley & Sons, Inc., p. 12.

Hayes and Wheelwright (1984b), p. 2.

Hayes and Wheelwright (1984c), pp. 20–40.

Hayes and Wheelwright (1984d), p. 83.

Imai, M. (1997). *Gemba Kaizen.* New York: McGraw Hill. *Gemba* means "real place"; the place where real things get done. According to the author, all levels of management should maintain close contact with Gemba, for it is the site of all improvement and the source of all information.

Juran, J. M. (1935). "Inspector's Errors in Quality Control." *Mechanical Engineering*, vol. 57, pp. 643–645.

Lamprecht, J. L. (1993). *Implementing the ISO 9000 Series.* New York: Marcel Dekker, pp. 59–66.

Liker, J. K. (2004). *The Toyota Way.* New York: McGraw-Hill, p. 280.

Madison, J. (1788). *Federalist No. 51*, February 8. The original quote is "If men were angels, no government would be necessary. If angels were to govern men, neither external nor internal controls on government would be necessary."

McClave, J. T. and Dietrich, F. H. (1988). *Statistics.* San Francisco, CA: Dellen Publishing, p. 364.

National Tooling and Machining Association (2011). NTMA Cleveland Chapter. http://www.ntmacleveland.org/. Accessed September 18, 2017.

Packard, V. (1960). *The Waste Makers.* London, UK: Longmans & Green, pp. 37–45.

Powell, J. H. (2016). "Recent Economic Developments, the Productive Potential of the Economy, and Monetary Policy." *The Peterson Institute for International Economics*, May 26. Data from the Bureau of Labor Statistics. https://www.federalreserve.gov/newsevents/speech/powell20160526a.htm. Accessed September 20, 2017.

Pyzdek, T. and Keller, P. A. (2003). *Quality Engineering Handbook.* New York: CRC Press. The authors quote Armand Feigenbaum from his book *Quality Control* (1951). New York: McGraw Hill, p. 125.

SAE Aerospace, SAE International Group (2004). *AS9102, Aerospace First Article Inspection Requirement.* Washington, DC: SAE International Group, p. 4.

Stimson, W. and Dlugopolski, T. (2007). "Financial Control and Quality." *Quality Progress*, 40, no. 5, pp. 26–31.

Thomson, W. (Lord Kelvin) (1891). *Popular Lectures and Addresses*, vol. 1, 2nd ed. London, UK: MacMillan.

Wright, G. (2006). Productivity Growth and the American Labor Market. In *The Global Economies in the 1990*, edited by Rhode, P. W. and Toliono, G.. Cambridge, UK: Cambridge University Press, Chapter 7, pp. 139–160. http://web.stanford.edu/~write/papers/2664_001.pdf. Accessed October 4, 2016.

17

Forensic Sampling of Internal Controls

In litigation concerning a performer's operations, the forensic systems analyst will turn to discovery to gather evidence of the effectiveness of operational processes. The evidence will be an historical record of references, records of activities, reports, and of various associated data including emails. This information is bounded in time by the epoch in litigation. With respect to the forensics of operational systems, that evidence is relevant that refers to or is related to the conformity of internal controls to recognized good business practices and to contract requirements.

Forensic Systems Engineering: Evaluating Operations by Discovery,
First Edition. William A. Stimson.
© 2018 John Wiley & Sons, Inc. Published 2018 by John Wiley & Sons, Inc.

Some formal means of extraction and inspection of the evidence may be required if the analysis is to have integrity. The formal means will follow established legal procedures of discovery, but not all the data may be examined. The data may be sampled, which is a well-defined discipline in statistics with its own established procedures.

Sampling is the process of taking samples from a *parent population*, and based upon the statistics of the samples, inferring an estimate of the properties of that population. *Population* is a statistical term for a set of objects under study. Forensic systems engineers and analysts are concerned with the population of the internal controls of operations and whether they are effective. They identify dysfunctional processes and determine whether its controls are the root causes of process failure.

The evidence that is sampled may take one of two paths. The first is that it is no longer treated as a sample but becomes the universe of interest—all that is important in the epoch of litigation. Litigation strategists may decide that the evidence stands by itself, with no attempt to project its parameters to the character of the parent population in discovery.

The alternate path is that the evidence may indeed be used as a sample from which statistical inference will be made to the population from which it was drawn. In this case, forensic systems analysts must follow formal sampling procedures if inferences are to be made from the observations. The integrity of the assessment will depend on the conformity of the sampling plan to recognize sampling procedures such as those described in Grant and Leavenworth (1988a). Sampling or not, the opinions of experts will be challenged and any departure from the rules of evidence will be attacked. If one samples the evidence, then sampling techniques effectively become included in the rules of evidence.

As no process is perfect, most dynamical operations will have a random component in the key characteristics provided by the process. Some of the variation will not conform to requirements, which of itself is nonjudgmental. The primary concern of the analyst will be in process stability and capability as discussed in Chapter 9. In this chapter, we are concerned with ensuring that systems analysis remains valid in the presence of sampling.

17.1 Populations

In law, discovery is held to be the entire body of evidence of interest to the litigation and usually stands by itself as sufficient for resolution. Indeed, to bring in additional evidence requires following rigorous legal procedures. Then it is reasonable to assume that the law regards discovery as the universe of material information for the case in dispute.

From a statistical perspective, too, the evidence in discovery can be regarded as the total population of interest. We are free to define our universe as we choose. To simplify matters, we shall align engineering concepts with those of the law and from the point of view of statistical analysis, consider that the evidence from discovery stands by itself. The analyst may use all of the evidence contained therein, or choose to sample from it. The decision to accept or reject the sample depends on the sampling rules, but will not reflect beyond the sample. Hence, an assessment by the analyst of the effectiveness of controls may be done by some form of sampling.

The evidence in discovery is not homogeneous and it is likely that the investigator will not be interested in all of it. What to select and how to select it are decisions that may be challenged in court, so prudence suggests that recognized and formal procedures be used in the selection process. If the forensic systems analyst does not have a background in sampling statistics, then the reader is encouraged to review Appendices C–F. This chapter is a forensic interpretation of those appendices.

17.1.1 Sample Population

Forensic systems engineering is concerned with the effectiveness of the operations in litigation. Invariably, this concern will focus on the existence and effectiveness of the internal controls of operational processes. The set of these internal controls then becomes the *sample population.* In this consideration, the controls are those policies, procedures, documents, and actions that the company employs to ensure the compliance of their operations to contract and legal requirements.

The analyst tests the controls to verify that they are functioning properly. The test is rather straightforward: if the controls conform to requirements, then they are assumed to work and are regarded as conforming; if the controls do not conform to requirements, then they are assumed not to work and are regarded as nonconforming. A nonconformity in the sampled population is often called a *deviation* because it deviates from the requirements.

The judgment of the effectiveness of the internal control system is made by comparing an estimate of the process deviation rate made from sampling, to an acceptable deviation rate established and agreed upon by the performer and consumer and which will be listed in the sampling plan. This approach is quite similar to an acceptance sampling scheme used in manufacturing and is universally acknowledged as proper manufacturing procedure. Therefore, the forensic systems analyst should use a similar approach.

17.1.2 Homogeneity

Homogeneity of a population is an important concept in statistical analysis. A homogeneous population is one in which all the data are similar to each other or have identical traits. The evidence contained in discovery is rarely homogeneous and this diversity of character must be taken into account in a sampling strategy.

For example, consider customer charges a performer with providing units of dubious quality, while demanding full price. Conceivably, three experts will be needed in the inquiry. A forensic systems analyst will be concerned with the effectiveness of the performer's system of operations. A financial analyst will be concerned with the payments demanded by the performer and the financial controls involved in the corporate financial system. The attorney leading the inquiry will act as the contract expert, concerned with contract requirements and whether false claims are appropriate. In this example of performance litigation, three separate but homogeneous samples were taken from discovery as each expert examined a sample of evidence from systems different in their nature. From the standpoint of statistical analysis, this approach is perfectly valid and is called *stratified sampling*.

The analysts must each strive to ensure that their sample is homogeneous. This identity is necessary to ensure that each member of the sample have the same variability. Consider the case of a sample of the corporate operations system taken from discovery. Although limited to operations, in general, even this sample will not be homogeneous but will include evidence of diverse controls concerning differing subsystems: documentation, management, resources, operations, supplier control, process stability, and verification and validation of product and process. Hence, it might well be that in most litigations there will be nested or double, stratified sampling.

Neither single nor nested strata are a problem in statistical estimations if statistical rules are strictly followed. Nor would they be in law. The assessments of the subsystems can be easily unified into an effective judgment by the lead attorney of total system performance if the subsystems are homogeneous.

There is a way to ensure homogeneity of diverse controls, which is explained in Appendix D and that can be implemented by sampling by *attribute*. Each evaluation is reduced to a yes or no decision. By defining all control assessments as binary decisions means that the entire control sample can be treated as homogeneous. At the same time, this tactic aids in providing a sufficiently large population for valid statistical inference. Attribute sampling is strongly recommended later in this chapter, as it offers many benefits to analysis, homogeneity being one of them.

17.1.3 Population Size

The size of a sample of evidence taken from discovery must meet certain statistical rules in order to make inferences about the total population from study of the sample. In a stratified sample of, say, operations, there would be different sample sizes for each stratum: documentation, supplier control, and so on. Sample sizes are usually chosen from the needs of the litigation strategy and from the experience of the attorneys and forensic witnesses. The greater the sample size, the higher the confidence level in the results.

The reader is directed to the section on sample size in Appendix E, and in particular to Table E.1 in order to gain some insight into choosing a sample size that will accommodate the various risks that are associated with sampling. If a sampling plan is adopted, then it is essential to use a recognized procedure for choosing the sample size. If the required sample size cannot be met, then non-statistical sampling is a viable alternative, as described in Appendix F.

17.1.4 One Hundred Percent Inspection

One hundred percent inspection means that there is no sampling—the entire available body of operational controls is tested. This eliminates the problem of sampling errors, but it can be shown that 100% inspection is only about 80% effective (Juran, 1935). Also, it is quite unlikely that the forensic systems analyst will have all of the evidence available in operations. While true that Defense and Plaintiff each have access to the entirety of discovery, as they compose their lists they will have differing interests and focus and will use different titles for each piece of evidence. Invariably, the analyst will be sampling, at least to a degree. As an aside, it can be quite embarrassing and possible damaging to be faced with a piece of evidence, available in discovery, that you did not know existed.

One hundred percent inspection is recommended in this book if it is possible to do so because the evidence in the sample will often stand by itself, making it unnecessary to infer to the total population. This is particularly true when the sampled evidence indicates systemic process failure. If misfeasance or malfeasance is firmly established within the sample, there is no need to go further. The converse may not be true; there may be no indication of misfeasance in the sample, making further pursuit probably unfruitful.

17.2 Sampling Plan

A sampling plan is a method of gathering and analysis of evidence. A formal approach to the extraction and inspection of data is required if the analysis is to withstand challenge. This formality can be met with a sampling plan of well-established procedures. There are two kinds of sampling plans: statistical and nonstatistical and each offers certain advantages. Moreover, both are recognized as admissible by the Institute of Internal Auditors (2013). Both types of plans are discussed at length in Appendices E and F, respectively, and the reader is encouraged to study both of them and certainly that appendix most pertinent to the reader's situation.

17.2.1 Objectives

When we start out on a new plan, there is often a tendency to simply swing into action. In statistical analysis, as in law, this is a dreadful tactic and I repeat the admonition of John Gibson (1990) relative to the tendency of programmers to "throw together a few lines of code, just to get started." This is a bottom-up strategy with ambiguous direction. Every plan must begin with an objective or set of objectives if you are to ever arrive at a designated goal, and establishing goals is a top-down activity.

In the investigation of systems in litigation, the first objective is to identify the processes in question and then to examine their performance for effectiveness. Invariably, this calls for a test of the internal controls. The parameters of interest will be those of process stability, capability, and nonconformity or deviation rate, because they reflect the corporate management's duty of care. The evaluation will be a simple attribute test of acceptable or unacceptable, depending on whether a control conforms or fails to conform to specifications or requirements.

Controls take a variety of forms so that the measurement might well involve metrics, parameters, methodology, validity, parametric range, sensitivity, statistical analysis, test specifications, instrumentation, and standards. Clearly, there is a lot more to measurement than just "doing it." The final objective is to verify that you know before you start what you are going to do and how you are going to do it.

17.2.2   Statistical and Nonstatistical Sampling

Statistical sampling applies the laws of probability to the sampling scheme to measure the sufficiency of the evidence gained, to evaluate the results, and to estimate sampling risks. It also allows you to make inferences of the properties of the sample to the lot from which it was taken. If sufficient data are available in discovery to conduct a statistical appraisal of effectiveness, then a statistical sampling plan may be the best approach. You have the rigor of a formal discipline in which to present the evidence.

Sometimes it is impractical or difficult to meet the requirements of statistical inference when sampling from discovery. For example, given the preselected basis and scope of discovery, randomness and adequate sample size may be difficult to achieve. Therefore, forensic investigators may prefer to conduct a nonstatistical plan.

A sampling plan is *nonstatistical* when it fails to meet the criteria required of a *statistical* sampling plan. Nonstatistical sampling plans rely primarily on subjective judgment derived from experience, prior knowledge, and current information to determine sample size and to evaluate results. A properly designed nonstatistical sampling plan can provide results just as effective as those from a statistical sampling plan, except that it cannot measure sampling risk. Therefore, the choice of a statistical or nonstatistical sampling plan depends upon the trade-off of costs and benefits relative to sampling risks.

At first blush it may appear that a nonstatistical plan is at best a second choice and in academic circles it could well be. With statistical sampling you can specify the degree of confidence in your results, say anywhere from 90 to 99%, and statisticians would understand that these are indices of probability and weigh them accordingly.

Juries might not. True, statistics is a powerful tool in the study of uncertainty because it offers measures, but it also offers complexities. It has been found in litigation that it is more difficult for a CPA firm to defend itself if it used statistical sampling rather than nonstatistical sampling (Fuerman, 2009). Suppose that you can say with 90% confidence that a particular control is effective (or ineffective, depending upon your role). Then what can the opposing counsel do with the 10% remaining? Will the jury understand the debate? Will they become confused between the argument of probabilities and the judge's instructions about reasonable doubt?

The selection of a sampling plan, or even whether sampling should be done at all, depends on the individual case. If there is an abundance of data and a decision is made in favor of a statistical strategy, then the sampling plan should be developed according to the discussion of Appendix E. If a nonstatistical sampling plan is chosen, then you should use procedures that are similar to those of statistical plans in order to enhance the integrity of your conclusions. Appendix F describes the approach to a nonstatistical strategy.

17.2.3 Fixed Size and Stop-or-Go

Of the many different types of statistical sampling plans, two of them are particularly adaptable to the partial audits taken from discovery. The first, *fixed size attribute sampling*, can accommodate developing systems where system deviation may be relatively high and easily measured. Fixed size would be appropriate in litigation of a contract of design and development or in which these phases of activity are a significant part of the overall contract.

The other is *stop-or-go sampling*, which can be used in stable systems with low deviation rates and hard-to-find errors. "Stop-or-go" is an attribute sampling strategy from the financial audit world and is used for controls whose expected deviation rate is very small. It is more efficient than a fixed-size plan because it uses small sampling numbers as long as the measured deviation rate is very low and in which the audited population is segmented into blocks. If the sample deviation rate remains small over a few blocks, the audit may stop. If the measured rate is higher than expected, the auditor has the option of increasing the sample size and continuing, or of stopping the audit of that control system.

Stop-or-go sampling is appropriate to controls of operations because of the natural grouping of operational functions, say documentation, production, resources, and so on. If the control system has a low deviation rate and if attribute testing is performed on each group, then an average system deviation rate can be easily determined, depending upon the litigant strategy. It may be preferred to retain individual group averages.

Whether the forensic team selects one or the other depends largely on the expected system deviation rate and the sample size possible from discovery. Fixed size can accommodate large errors, but needs larger samples. Stop-or-go is easily adaptable to evidence grouped in strata, but it is designed for small system errors. Appendix E provides guidelines on both methods and the analyst may find that a plan combining both methods is the most effective.

17.2.4 Sample Selection and Size

We have defined the entirety of evidence in discovery as representing systems of various nature. The forensic systems analyst is concerned with the evidence relevant to operations and this population may stand by itself. We can conduct 100% inspection or we can sample. If we choose to sample, then a few common methods of doing so are random, systematic, and haphazard sampling. These methods are described in Appendix E, as well as are the various methods of determining sample size.

It may be that a "smoking gun" will be found in discovery that provides all the information necessary to a firm evaluation of the controls of operations and that renders further sampling unnecessary. However, that will not be clear until you have seen the evidence and you should not see the evidence until you have chosen a sample selection and size that can withstand challenge.

17.3 Attribute Sampling

The performance of a system depends upon its stability and capability. In Chapter 15, we learned that with a system in control, process instability should be a rare event. Therefore, processes have controls to ensure their effectiveness. We learned also that if a control conforms to quality management system (QMS) requirements, then it is assumed to be effective and the controlled process to be providing units in conformance to specifications.

Attribute sampling is a statistical method of sampling from populations that is used with tests of controls. It allows the auditor to verify the conformity of the control or lack thereof to a specified requirement. Hence, attribute sampling is entirely appropriate to the testing of controls in forensic systems analysis.

17.3.1 Internal Control Sampling

In Chapter 5, we defined an internal control as a process to assure effectiveness of operations, reliable records and reports, and compliance with standards. In the earlier section on homogeneity, we established that a control can be viewed as an attribute—it either works or it does not, and thereby we verify the compliance of the control. Therefore, the analyst must be clear on what is essential in verifying a control. For example, verification and validation processes are controls. You are primarily interested in whether the controls were conducted according to standard procedures.

As a detailed example, assume that you are examining a company's receipt inspection process which objective is to prevent unverified parts from being used in production. The receipt inspection process is a control. Relevant procedures define the process in six steps:

1) Receiving. The parts are received at a point designated as the receiving station.
2) Holding. An operator labels the parts "not inspected," and then puts them in a quarantine area until they can be tested.
3) Inspection. An inspector inspects or tests the parts for conformance to the purchase order, according to a specified acceptance scheme.
4) Labeling. The inspector labels the good parts "Inspected OK," and labels the bad parts as "Inspected and unacceptable: return to vendor."
5) Recording. The inspector records the measurement results as a quality record, sending a copy to Purchasing.
6) Traffic. A forklift operator moves the good parts to inventory or to a specified shipping dock and moves the rejected parts to an assigned area for repair, scrap, or regrade.

(*Regrade* means that the product is reclassified for other purposes. For example, the maker of fabric may require zero defects per yard in its product. If there are one or two defects per yard, the fabric may be sold as remnants; if there are three or more, it may be sold as rags.)

All of these steps are part of the control process and your task as a systems analyst is to verify that the steps described in the procedures were being followed according to defined and industry-accepted procedures. The actual rejection rate of the received parts is less of a concern so long as the rejected parts were not provided to the customer.

So how do you evaluate this control? Suppose that the data show that one of the steps was not followed, but the rest were good. Whether you assess this receipt inspection as a nonconforming process depends on the severity of assessment.

The systems assessment is a go/no-go decision no matter the nature of the control because by defining all control assessments as yes or no decisions means that the entire QMS control population can be treated as homogeneous (Mills, 1989). This strategy can enhance statistical inference by providing a sufficiently large population.

Hence, controls are sampled and tested to determine if they conform to stated requirements. If they conform, then they are deemed effective and acceptable. If they do not conform, then they are deemed ineffective and are rejected. This is a binary decision. The record of the number of units conforming or not conforming to some requirements is called count or attribute data. Attribute sampling is the sampling of count data.

Financial auditors use attribute sampling extensively in the testing of financial controls. Appendix D describes the process in this way: A financial auditor tests controls to see if they are functioning properly by measuring compliance with the control procedures in terms of deviation rates. Compliance in a given test means the control works. Noncompliance means that the control does not work—a deviation exists.

The financial test of controls is quite similar to an acceptance sampling scheme used in manufacturing. As forensic systems analysts will also use the same decision process in tests of controls in operations, then attribute sampling is most appropriate. The attributes may be expressed as a fraction rejected: the ratio of the number of rejections to the total per sample. The reader is encouraged to study Appendix D on this matter.

17.3.2 Deviation Rates

In attribute sampling, departures from conformity are viewed as deviations from the norm. There are two such rates: the acceptable deviation rate and the system deviation rate.

17.3.2.1 Acceptable Deviation Rate

Systems analysts concede that variation exists in any scheme of production and service. In any given phase of performance, a stable rate of variation will exist, some of which will be nonconforming. Management will implicitly define an acceptable rate, similar to an acceptance number in acceptance sampling. In the fiscal audit profession, this is usually called the acceptable deviation rate (ADR). In this book I adopt "acceptable deviation rate," yielding to the seniority of fiscal auditors. As there will always be a system deviation rate, the forensic analyst must define and work with an ADR, perhaps by assigning an index of severity to the deviations as discussed in Chapter 14.

In my experience, a good strategy is to pass over any random deviation and focus on systemic failures. Why? Suppose that the sample of evidence indicates what appears to you as a rather high rate of random deviation. You can assemble a team of statistics experts to testify to the validity of your findings. The opposition will assemble a team of experts equal in standing to declare that your findings are without merit. The ensuing technical argument will be incomprehensible to the jury and whatever its decision, it may not be based on the merits of the argument. On the other hand, if a control simply does not work or is missing, then systemic failure is self-evident with consequences described in Chapters 13–15, along with a very strong case for damages.

If you conduct an audit in a given area of inquiry and find zero nonconformities, it does not mean that there are none, it simply means that you did not find them. Conversely, if you find several nonconformities, it does not mean that the system of controls is ineffective. That depends on the amount and nature of variation and its likely impact to the customer. On the other hand, systemic failure indicates an unstable and incapable process with almost certain strong negative customer impact.

17.3.2.2 System Deviation Rate

In this book, the terms "system" and "process" are equivalent. Moreover, whether any given entity is a system, subsystem, or element in a system is a matter of perspective. The term "system deviation rate" as used here can refer to the deviation rate of the QMS or to any subset of it, including the block of "processes" that is being investigated from the subject population.

In forensic examination, we usually do not know the system deviation rate, but must estimate it from data in discovery. In principle, the analyst will compare the estimated system deviation rate to the acceptable deviation rate. If the estimated rate is lower than the acceptable rate, subject to appropriate confidence level and sampling risk, then the control procedures are deemed effective. Otherwise, they are ineffective.

I say "in principle" because conventional sampling theory is appropriate in litigation only if there is a sufficiency of evidence supporting inference. The sufficiency depends on the underlying distribution of the process deviation. Hence, if it is reasonable to assume that the distribution is mound-shaped, then a sufficiency would be a minimum of 30 items of data from the same population (McClave & Dietrich, 1988).

If nothing is known of the underlying distribution, or if the available evidence in discovery offers only small sample inference, then there are two alternative strategies: (i) to use a nonparametric method or (ii) to forego system deviation rates and pursue the issue of systemic failure.

17.3.3 Sampling Risks

The best sampling plan in the world can provide only an estimate of system performance. The statistical deviation rate is only an estimate of the system's true deviation rate, and the statistical variance is only an estimate of system variance. Yet, the forensic analyst must make a decision about the system based on these estimates. There is a certain risk that the estimate will be wrong, one way or the other. A risk is a probability of loss, often associated with a decision. In this sense, there are three risks in sampling: control risk, alpha risk, and beta risk.

17.3.3.1 Control Risk

A control risk is the risk that a given control structure may fail to do its job. Given that control systems in operations are designed by engineers, a control that fails to do its job is more likely to be due to an impromptu or expedient change in the control structure than from its design. Indeed, in my experience the overwhelming majority of control failures are due to arbitrary changes in control structure and have nothing to do with sampling techniques. Such controls should be regarded as deviations from compliance and accounted as such.

17.3.3.2 Alpha and Beta Risks

There are two risks in the sampling itself, known as *alpha* and *beta*. The alpha risk is that the sample will indicate that an effective control is ineffective. The beta risk is that the sample will indicate that an ineffective control is effective.

The analyst may make an *alpha error* by deciding, on the basis of a sample, that the control was ineffective and conclude that there was a high risk that nonconforming product had been delivered to the customer. Yet in reality the control was effective, with little nonconforming product. Thus, the analyst will introduce misleading evidence, create increased and costly litigation, and possible wrongful conviction, to correct a problem that never existed.

A *beta error* is made if the analyst decides, on the basis of a sample, that the control was very effective, with low risk of nonconforming product. Yet in reality the control was ineffective, providing an unacceptable level of nonconforming product. Thus, in making a beta error, the analyst is nonresponsive to an unacceptable rate of nonconforming product that was delivered to the customer.

Both alpha and beta risks can be predetermined with various sampling strategies such as varying the sample size. The alpha risk reflects the level of significance and confidence desired in a test; the beta error is inversely related to the sample size—a larger size provides a lesser risk. Also, the two risks are inversely proportional. The reader is encouraged to read Appendix D in the matter of alpha and beta risks, and their similar probabilities, consumer risk, producer risk, type I risk, and type II risk. The forensic systems analyst will come upon each of these titles, depending on the source of documents and the context, which itself is interesting. At the end of the day, they all have to do with the producer–customer relationship and how manufacturing and service risks are distributed between them.

17.3.4 Confidence Level

In statistics, *significance* does not refer to the importance of an observation but to the probability, in taking a sample, that the observation is characteristic of the population and not due to sampling error. Thus, $\alpha = 0.05$ means we can distinguish designed effect from error down to 5%. The level of significance should be chosen before the test, as discussed in Appendices D and E.

It follows that the confidence that we can have in the measurement is defined by a *confidence coefficient*, $Cc = 1 - \alpha$. When the confidence coefficient is expressed as a percentage, it is called the *confidence level*. Thus, if $\alpha = 0.05$, then $Cc = 0.95$ and we have a confidence level of 95% that the test will yield a valid estimate.

17.3.5 Evaluation

Whatever method of sampling is chosen, the approach to evaluating the result of the test is the same: a comparison of the measured deviation rate of a process to an acceptable deviation rate. However, in forensic systems analysis this comparison is not direct, for the sample is simply one outcome of a process with natural variability. Another sample would yield a different deviation rate. Therefore, the best approach is to determine on the basis of the sample mean, what would be the highest process deviation rate within a given level of confidence?

In the case of statistical sampling, you first estimate the system deviation rate from historical data, pilot data, or knowledge of the process under test. This estimate will affect the decision on the required sample size. After sampling, you compare the estimate with the measured rate. Then as described in Appendix C, you can determine from software such as EXCEL, the highest likely system deviation rate. You then compare this highest probable deviation rate to the acceptable deviation rate to assess whether the control is conforming or nonconforming.

In the case of nonstatistical sampling, you cannot make a reliable estimate of sampling risk, but you must make a judgment on whether the difference between the acceptable deviation rate and the measured deviation rate is an adequate allowance for sampling error. For example, suppose that the measured deviation rate was significantly greater than the expected rate. Thus, the bounds and confidence levels of a test of controls are critical and the right questions must be asked. Appendix F describes a comprehensive approach to nonstatistical sampling.

17.4 Forensic System Caveats

It is comfortable to have the power of statistics behind you because "power" suggests strength and authority. However, the field of statistics has its own definition of power that is more appropriate to forensics. The power of a statistical test is defined as the ability of a test to detect an effect if there is one. Specifically, power $= 1 - \beta$, where β is a type II error—the failure to detect a change in the system. Type II error is inversely proportional to sample size, so the first criterion in deciding whether to use statistical or nonstatistical sampling, or to sample at all, depends on the availability of evidence in discovery.

Whether to use the fixed size or stop-or-go approach depends upon the expected and measured system deviation rates. If the contract includes development, you can expect rather large deviation rates and "rather large" is a vague guideline. You can easily be drawn into technical arguments that can have unpredictable responses from a jury.

I favor the systemic failure approach described in Chapters 14 and 15, if such failure exists, because each result stands by itself. Misfeasance is clear and if the nonconformity is sustained, the risk of product nonconformity approaches unity and β approaches zero. However, whichever approach is taken, reliance must be placed on two characteristics essential to forensic systems analysis: professional judgment and skepticism (International Federation of Accountants, 2009):

1) *Professional judgment* is defined as the application of relevant training, knowledge, and experience within the context provided by auditing and ethical standards, in making informed decisions about the courses of action that are appropriate in the circumstances of the litigation.
2) *Professional skepticism* is defined as an attitude that includes a questioning mind, being alert to conditions which may indicate possible misfeasance due to error or fraud, and a critical assessment of audit evidence.

Skepticism includes being alert to the following:

- Audit evidence that contradicts other audit evidence obtained.
- Information that brings into question the reliability of documents and responses to inquiries to be used as audit evidence.
- Conditions that may indicate possible fraud.
- Circumstances that suggest the need for additional audit procedures.

References

Fuerman, R. D. (2009). Associate Professor, Sawyer Business School, Suffolk University, Boston, MA. The cited comment is from his accounting course and is unpublished. Professor Fuerman may be contacted at: rfuerman@suffolk.edu.

Gibson, J. E. (1990). *How To Do Systems Analysis*. Charlottesville, VA: School of Engineering, University of Virginia Workbook, p. 43.

Grant, E. L. and Leavenworth, R. S. (1988a). *Statistical Quality Control* (Part Two). New York: McGraw-Hill.

Grant and Leavenworth(1988b), p. 60.

Institute of Internal Auditors (2013). *Practice Advisory 2320-3: Audit Sampling*, Issued May, p. 2. https://www.iia.nl/SiteFiles/PA_2320-3%20(1).pdf. Accessed September 13, 2016.

International Federation of Accountants (2009). "Overall Objectives of the Independent Auditor and the Conduct of an Audit in Accordance with International Standard on Auditing." *International Standard on Auditing 200*. http://www.ifac.org/system/files/downloads/a008-2010-iaasb-handbook-isa-200.pdf. Accessed September 17, 2017.

Juran, J. M. (1935). "Inspector's Errors in Quality Control." *Mechanical Engineering*, vol. 57, pp. 643–644.

McClave, J. T. and Dietrich, F. H. (1988). *Statistics*. San Francisco, CA: Dellen Publishing, p. 364.

Mills, C. A. (1989). *The Quality Audit: A Management Evaluation Tool*. New York: McGraw-Hill, p. 173.

18

Forensic Analysis of Supplier Control

Companies that make things or that provide services for sale invariably require supplies from sources external to themselves. A manufacturer of metal cabinets may purchase sheet metal from a steel manufacturer; a shoemaker purchases leather; a baker purchases flour; and a bank purchases forms and other kinds of software. Sometimes external services, too, are purchased, financial services, for example. The external sources are called by a number of names, among them "suppliers," "vendors," or "subcontractors." All of these names are used here and there in this book, but for brevity in this chapter, we shall call an external source a *supplier*, in keeping with ISO 9000 (ANSI/ISO/ASQ, 2005), which defines a supplier as an entity that provides a product or service to an organization.

Forensic Systems Engineering: Evaluating Operations by Discovery,
First Edition. William A. Stimson.
© 2018 John Wiley & Sons, Inc. Published 2018 by John Wiley & Sons, Inc.

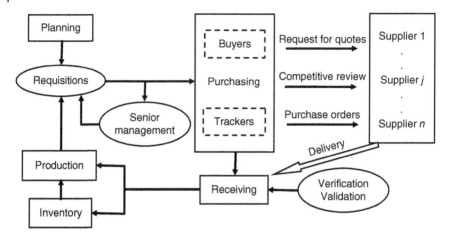

Figure 18.1 A general purchasing system with supplier control.

Good business practice requires that a company verify and validate the quality of supplied resources used in creating its products or services before those resources are integrated into the final units. The integrity of the supply system is safeguarded by controls to ensure that purchased units conform to specifications, provide for contractor assessment, track the status of purchased data, and allow for customer input during the purchasing process. The necessary controls are determined by the company, often in meeting the requirements of a standard of performance. In a later section of this chapter, commonly accepted requirements of supplier control are cited as principles, along with descriptions of appropriate controls. The nature of these controls indicates that they are effectively functions of verification and validation of externally provided units.

A closed loop purchasing system is shown in Figure 18.1 and contains the various activities just described. The system is robust because it converges inventory to dynamic demand, at the same time providing control of the quality factors. Note that in this control loop, the external suppliers are part of the purchasing system and the circuit is closed by verification and validation at receiving.

Figure 18.1 may seem complicated at first glance, so let us break it down to its basic components. The basic purchasing process has three stages:

1) Requisition
2) Receiving
3) Management of purchased units.

The requisition process should include procedures for matching customer requirements to the purchase of parts and materials; a solicitation process of inquiry, quotation, and tender; a process for selection and evaluation of suppliers; timely order and delivery of resources; an effective purchase order format; and long-lead time parts ordering policies and procedures.

The receiving process should include verification and validation of units and materials against the purchase order and the customer requirements and supplier evaluation in terms of effectiveness and efficiency.

The unit's management process should include protection and care of received material, timely delivery to the end user, an inventory mechanism with triggers for maintaining material of varying inventory turnover.

From a systems viewpoint, whether all the elements of the purchasing system are in-house, distributed, or even composed of other companies is immaterial. They are part of the system. However, having a supplier 5000 miles away is not the same as having that supplier next door, as is evident when the dynamics of distribution and transportation are taken into account in the analysis of a purchasing system. If the evidence indicates that the verification and validation controls of supplied units are inadequate or none at all, then the purchasing system is operating in an open loop and the forensic analyst must conclude that there is no control. In Chapter 11, it is shown that the controllability of an open loop system ranges from problematic to nonexistent, depending on the period of time in which the loop is open.

18.1 Outsourcing

In Chapter 7, under the heading of resource management, I said that in the global economy much of the manufacturing task is distributed by contractors to numerous suppliers as subcontractors, which diffusion greatly increases the difficulty of supplier control. I listed several supply chain problems that seem to recur in litigation: ghost inventory, unmonitored outsourcing, substandard purchased units, and ineffective flow-down. Indeed, where the outsourcing is widespread, supplier control may compete with verification and validation as a litigious activity. To better understand the difficulty in effective supplier control, the forensic systems analyst should be familiar with the nature of resource management as it is practiced today.

Previous examples of supplies described the case where a company might purchase parts or services from an external source because the company itself does not provide the needed resource. An automobile manufacturer does not make tires, although tires are essential to the operation of its automobiles. However, sometimes companies may lease out certain of its business functions or services to an external source that they formerly made themselves. This practice is called *outsourcing*.

Peter Drucker's quote "Do what you do best and outsource the rest" (Vitasek, 2010) is regarded as a major impetus of the large outsourcing phenomenon that began in the early 1990s and continues today. The Drucker influence is indicated by this statistic: an average US manufacturer outsources 70–80% of its finished product (Corbett, 2004). One suspects that although Drucker did not recommend outsourcing core competencies, they are nevertheless outsourced to a significant extent. The magnitude of outsourcing is indicated by a comment of Robert Hoyer (2001) that "Ford no longer makes cars—they assemble them." Hoyer's implication is that almost everything in your automobile is made elsewhere, some of it offshore.

Offshore outsourcing refers to leasing out functions to overseas suppliers and has increased markedly in the twenty-first century, as witnessed by the plentitude of products that carry the name of an American producer but are made overseas. Lach (2017) reports that US multinationals shifted millions of jobs overseas in the 2000s. Data show that the big brand-name companies that employ one-fifth of all American workers cut their work forces in the United States by 2.9 million during the 2000s while increasing employment overseas by 2.4 million. Lach offers a further example: the global electronics contract manufacturing industry reached a staggering $360 billion of revenue in 2011 and is expected to expand to $426 billion within a few years from that time. Much of this increase derives from American firms contracting manufacturing and service offshore. My purpose here is not to comment on the effect, good or bad, of outsourcing offshore but to indicate the inherent complexity of supplier control in doing so.

While Peter Drucker enjoyed significant influence, he was simply offering a solution to problems that many companies had, hence their willingness to follow his advice. The following are some of the reasons why companies choose to outsource certain activities (Pine, 2017):

- Reducing and controlling operating costs
- Improving company focus
- Gaining access to world-class capabilities
- Freeing internal resources for other purposes
- Streamlining or increasing efficiency for time-consuming functions
- Maximizing use of external resources
- Sharing risks with a partner company

It might seem that supplier control is simply one of many controls in business and should be no more trouble than any of the others, but this idea breaks down at some point when there is an increasing reliance on outsourcing. Why? Intuitively, we might conclude that the more suppliers, the more complex the problem of managing the activities of purchasing, scheduling, transportation, inspection, quality, inventory, and networking. There are many ways to measure complexity, but one of the metrics of organizational complexity is the degree of organization. Within that context, I postulate that complexity can be measured by the number of the system control loops. Therefore, increasing the number of suppliers increases system complexity because each supplier requires numerous controls. Moreover, as one control loop may feed another, the increase is not linear and may be exponential. However, complexity is not a useful metric in forensics because while it indicates the difficulty of system operation, it does not measure effectiveness per se. Yet, system complexity can serve as a warning flag to the forensic analyst and may be considered a contributing factor in the search for root cause.

Mathematical models exist for optimizing the design of supplier controls, but the forensic problem is focused on that which has already happened. Given the geographic structure of the purchasing functions, it is clear that operational complexity increases with increasing distance from supplier to performer and with increasing numbers of suppliers.

From a forensic systems point of view, whether a company purchases resources from an external source because it does not and cannot provide them itself, or because the company chooses to outsource some of its own activities is irrelevant. The principles of supplier control remain the same and the supply chain must be managed accordingly.

18.2 Supply Chain Management

The activity of managing multiple sources is called *supply chain management* (SCM). Although outsourcing and supply chain are different names for overlapping activities, each is a field of study in its own right and the forensic analyst should be familiar with their distinctive characteristics.

The basis of supply chain management (SCM) is the realization that most units that comes to market result from the efforts of many organizations whose efforts must be coordinated. SCM is the management of materials, information, and finances as they move from supplier to manufacturer to dealer to consumer.

Materials management concerns the movement of goods from a supplier to a customer. The information flow transmits orders and tracks the status of delivery. The financial flow concerns credits and payments. The purpose of SCM is to design and implement effective and efficient supply chains, from production to product development to the information systems needed to direct them.

The SCM system attempts to centrally control and unify the production, shipment, and distribution of a product. By keeping tight control of inventories, production, distribution, sales, and the inventories of company vendors, companies are able to cut excess costs and deliver products to the consumer more effectively.

Each company that utilizes SCM designs its own system, some more comprehensive than others. Forensic systems strategy, too, will vary from attorney to attorney and in this book, the systems frequently found in litigation are categorized as discussed in Chapter 7: management, supplier control, verification and validation, and documentation. Hence, the forensic systems analyst studies those aspects of SCM that focus on these areas. The financial concerns in litigation related to suppliers will usually be assigned to another forensic team with expertise in financial accounting.

18.3 Forensic Analysis of Supply Systems

18.3.1 Basic Principles of Supplier Control

One of the unexpected but interesting features of forensic systems engineering is how much the engineering approach adapts to the strategy used by attorneys. English common law is codified to a certain extent, but to a great degree it draws on precedent—what were the earlier judicial decisions in similar cases, and how well does the present case fit the earlier ones?

Similarly, the forensic systems analyst will draw on most of the rules, regulations, statutes, and standards pertaining to effective supplier control. However, these documents are rarely in a single body or even in a single library. Which ones are relevant and where can they be found? It will be up to the forensic systems analyst to find them. As an example, if one of the controlling standards is ISO 9001 (ANSI/ISO/ASQ, 2015), the supplier requirements will be listed in varying clauses, themselves with varying titles and wording depending on the version. The principles will be there, but a significant amount of time and thought will be needed in perusal of a large amount of documentation. Therefore, we list the principles of supplier control without citing clause numbers, relying on the analyst to learn what to look for in reviewing the evidence of operations performance. The major principles can be summarized as follows:

- Ensure that purchased units conform to contracted specifications
- Maintain a process of supplier assessment and selection
- Track purchasing status and data throughout the production cycle
- Establish effective customer relations
- Maintain procedures for V&V and storage of purchased resources
- Provide identification and traceability of parts, where required by contract or custom

The nature of these supplier controls indicate that they are, effectively, an extension of the verification and validation function applied to external provision.

18.3.2 The Forensic Challenge

Standards of performance tend to allow significant leeway in the implementation of requirements because of the many varieties of business structure that are expected to adapt to the standards. Although the original intention of outsourcing was to outsource support functions, today it is quite common and even acceptable to outsource core competencies. ISO 9001 (ANSI/ISO/ASQ, 2015) is no exception and forensic systems analysts must avoid a "by the numbers" approach to their considerations. There is seldom a "right way" to do something. The acid test of whether a control is effective is determined by the quality of the process results relative to customer requirements and not by whether the control fits some classic format. The final arbiter of any litigation is the jury, which will be focused only on whether the plaintiff has suffered harm.

Nevertheless, the analyst must determine whether a given control exists and whether it is effective. "You can lead a horse to water, but you cannot make it drink." This old saying can be applied just as well to the relationship of a performance standard to those who might consider it an imposition. Ineffective procedures are the bane of supplier control and very often come about because the responsibility of a company for the activity of its suppliers is often not emphatic in many standards of performance. The forensic systems analyst may find it necessary to refer to other requirements invoked in a given contract in order to establish that a company is always responsible for the quality of its delivered units.

18.3.2.1 Ensure that Purchased Units Conform to Contracted Specifications

This is a blanket requirement that covers the spectrum of supplier control and so may be perceived as too general to be effective. Granted that it offers no details of control, yet its generality is critical when taken literally. Ensure means ensure. The buck stops here. The company is responsible for the quality of final product or service irrespective of any agreements, accords, or licenses contracted to a supplier.

Very often, the controls needed to meet this principle can be found in many standards and most of them can be implied by an experienced forensic analyst: competent suppliers, appropriate skills and equipment, sufficient communication and information transfer, and effective verification and validation of supplied product.

Usually, the company will choose where the verification of parts conformance will take place and by whom. The verification function may be at the supplier's facility and conducted either by supplier personnel or by team from the company itself. Alternatively, the function could take place at the company facility, but if so there will almost never be redundant verification. If the verification function is assigned to the supplier, the responsibility for conformance remains with the company as prime contractor because this is the only entity recognized by the customer. The purchasing system is at risk if the supplier's quality management system (QMS) is independent of the company's QMS.

Chapter 7 of this book contains a discussion on flow down that should be repeated here. A supplier must receive timely and accurate information on the task to be achieved, an idea that is easy to understand but not so easy to ensure. Who informs the supplier of, say, critical specifications of an item to be fabricated? Is it Purchasing? Production? Design? Does the company use a procedure to ensure the timing and path of flow down? In a large organization, flow down can fall between the cracks and the mass production of an inferior part is not always the fault of the supplier.

18.3.2.2 Assessment of the Supplier Process

Ensuring that supplier parts and services conform to requirements is necessarily an after-the-fact activity: the assurance comes about in reviewing what has been done. But the company also has before-the-fact responsibility: ensuring that suppliers have the capability to meet the requirements before they are brought aboard. Thus, forensic analysis includes examining the selection and evaluation processes used by a company in choosing its suppliers.

In this examination, the analyst has no preconceived template; the company can choose any methods of selection and evaluation that it deems sufficient. For example, a company can "grandfather" its long-time suppliers, but having done this, all suppliers must thereafter be continually evaluated. However, given a selection process, the analyst must determine whether it is effective. A stream of consistently nonconforming resources implies an incapable supplier and raises questions about how that supplier was chosen. If the selection process seems inadequate, counsel may well consider whether the process itself is in violation of contract requirements.

I once reviewed a selection process in which supplier candidates provided proposals reflecting their capability and costs. The evaluation system seemed proper in that candidates were assigned scores, but somehow the supplier with the highest score was not selected. Rather, the winner of the bidding was a supplier whose proposal was rather noncompetitive. As that supplier was listed as a defendant in the complaint because of consistent provision of nonconforming parts, the selection process of the prime contractor was challenged.

Therefore, when there seems a consistent trail of poor supplier performance, the forensic analyst asks two questions: "What is the policy for choosing suppliers?" "How are suppliers evaluated?" Then in discussion with counsel, the analyst can contribute a technical assessment to an effective strategy of litigation concerning the supplier system.

18.3.2.3 Tracking

The purchasing system shown in Figure 18.1 includes a block entitled *trackers*. Tracking refers to the process of following the status of resources on order and obviously, a tracker is the person doing that job. In a business in which supply is in large volume, tracking is a full-time job and may include several trackers. For example, automobile manufacturing and ship building and repair are two industries with very large volume supply chains.

Suppliers have their own schedule, which may not conform well to that of the prime contractor. Long lead time items may take even longer than anticipated. An error in the purchase order, or a misreading of it, may result in a nonconforming delivery and an adverse impact on the critical path. Good tracking provides both formative and reactive responses to the supplier system, so that if discovery indicates a consistently poor supplier dynamic, it is reasonable to question the company tracking process.

As an aside, I have observed that tracking is a highly stressful job. You cannot take no for an answer but must plead, entreat, cajole, and argue because often millions of dollars are at stake in a threatened schedule. In my view, trackers are unsung heroes that have rescued many a contract.

18.3.2.4 Customer Relations

The kind of businesses that are likely to invite systems litigation are often those having an extended period of interaction between customer and performer. Customer relations are far more complex than in simply buying and selling because the dynamic picture can change from day to day and very often the customer will be, or should be, involved in the making of decisions. There are three essential components in this relationship: customer property, communication, and mutual satisfaction.

Within the context of supplier control, customer relations imply a three-way network: customer to company to suppliers. Granted that there may be no contractual relationship between customer and supplier, there is often a real-time relationship between them and in any case, the group forms an operational system and each performer in that system should be aware of the day-to-day dynamics. It is the responsibility of the company to ensure that its suppliers are aware of customer expectations and the intended use of the final product or service. Suppliers are part of the team.

Communication refers to the exchange of information relating to parts and services within the team, inquiries and change orders, customer feedback, handling of customer property, and agreement on contingency actions.

Customer property refers to any product, part or parts, information, or intellectual property that is provided *by the customer* to be used in the manufacture of a product to be delivered to that customer. Customer provision of property is quite common in some industries. For example, a company that replicates compact disks will receive audio tapes from its customers, replicate the music onto the disks, and then return the tapes to the customer when it delivers the final product—the modified disks. In a somewhat different scenario, the US government will deliver weapons systems to a shipyard to be installed on a newly constructed ship. Upon delivery, the government receives the ship with its own property installed aboard. In each case, the tapes and weapons systems always belonged to their respective customers. They never at any time belonged to the performing organization.

There are two responsibilities to customer property: stewardship and quality control. As steward, the company is responsible to safeguard the customer's property as inventory until such time as it is put into production. At that time the material falls into the mainstream of the company's quality control processes. Safeguarding means that the material must not only be protected from theft but also from the weather or damage on the floor or in the yard. The forensic analyst will be concerned with written and implemented procedures on how this stewardship is achieved.

As the quality of the customer property will impact the quality of the final units, the company is responsible for recording and reporting to the customer the event of lost, damaged, or delivered property unfit for use. This report is a quality record because the customer will deliver its property to the performer in good faith and assume that its responsibility has been fulfilled. If the property is somehow unfit, there may be grounds for dispute of contract and customer or performer grievance.

Customer property is received as any other material—through receipt inspection. The delivery date, condition, quantity, and fitness of the property are recorded and compared to contractual requirements. The property will then go to inventory until it is required in operations. It should be distinguished from other inventory by labels and location because it does not belong to the company; it belongs to the customer. The holding area need not be unique—other material can be in the general storage area—but it should be sufficiently isolated as to be easily distinguished. Of course, it must be safe from damage or misuse.

The material is integrated into the final product according to the job order, but this final product may not be called for immediately. If not, it again returns for holding, and again it has a unique status. The analyst should be aware of a possible ambiguity here. If the customer property is integrated into a final product, as in the case of the weapon system installed on a ship, then the performer has added sufficient value that routine care of final product is such as to also ensure care of the customer property therein. On the other hand, in the case of CD replication, only the intellectual property was removed from the tapes, which remain customer property. Not being in the stream of final product, the tapes require special care throughout the period of their retention. At some point the finished disks will be called for delivery and will transit in accordance with standard operating procedures.

Implementing the controls are the prerogative of the company; the forensic analyst is concerned only that the control works. For example, in some contracts an in-process unit may be the property of the customer even though it is being worked on by the company. Suppose that this property is to be integrated into a final unit but is unfit or damaged. This event should be reported to the customer for a decision on rejection or acceptance by concession.

If the forensic analyst finds evidence that a unit damaged in production was never reported to the customer, nor was there evidence of subsequent unit verification, then any later claim of acceptance by concession cannot be true. There are at least two contract violations here and further investigation is called for. Depending on the function of the unit that is nonconforming, or on the process that made it, there may be more, perhaps many more, nonconformities resulting from the first.

18.3.2.5 Verification and Storage of Supplies

Many performance standards, including ISO 9001 (ANSI/ISO/ASQ, 2015), require an organization to establish and implement the inspection or other activities necessary for ensuring that purchased product meets specified purchase requirements. Where the organization or its customer intends to perform verification at the supplier's premises, the organization shall state the intended verification arrangements and method of product release in the purchasing information.

Purchased product can be verified either at the supplier facility or upon arrival at the company facility, or both. The arrangement should be formal, described in policies and procedures and in the contract with the supplier. Verification at the supplier facility places most of the weight of quality on the supplier. The company can accomplish this verification in two ways. Company buyers or quality experts can visit the supplier facility and conduct an audit of its inspection procedures. Alternatively, the company may accept, with the delivery of the purchased product, associated control chart results, other statistical evidence of product quality, third-party certification of test results, and certifications of conformance from the supplier. Thus, the company can rely primarily upon the supplier's quality control procedures.

However, verification of product quality by the supplier does not relieve the company of its responsibility to provide acceptable product to its customers. The company need not repeat the verification process in its own plant, and major corporations often use supplier verification if they have hundreds of suppliers and the sheer volume of receipt inspection is inefficient or costly. This option is entirely reasonable, but there is inherent risk because, in practice, there may be a divergence of supplier and company QMS, resulting in a stream of unobserved nonconforming units. If the supplied units have a critical role in the final product, the company may choose to inspect them in its own facility. If so, there must be a formal procedure to notify the supplier of the outcome of such inspections.

Some customers may wish to verify the quality of purchased units themselves either at the company facility or perhaps at the supplier facility. This option will of course be in its contract with the company. The US government is often such a customer. ISO 9001 (ANSI/ISO/ASQ, 2015), as one example, makes it clear that inspection of purchased units by the customer absolves neither the supplier nor the company of its responsibility to the quality of its delivered units.

In some cases, supplied product cannot be verified by subsequent monitoring or measurement and as a consequence, deficiencies become apparent only after the product is in use or the service has been delivered. In such circumstances, the company should validate any processes for the provision of production and service where the validation shall demonstrate the ability of these processes to achieve planned results and contract requirements. The company should have established controls for these processes including, as applicable:

- defined criteria for review and approval of the processes
- approval of equipment and qualification of personnel
- use of specific methods and procedures
- requirements for records

If the verified and validated purchased unit is not put immediately into an assembly system, it will be placed in inventory until it is needed. Inventory is an important issue in supplier control and depends upon company operations strategy—push or pull production. During the audit of an operational system, I was always fascinated by the degree of in-process inventory. As discussed in Chapter 7, Japanese industry prefer just-in-time inventory and in current American industry, tight control of inventory is one of the goals of SCM. However, the United States is a vast nation and tight control is quite difficult. And, somewhat surprising to me, political issues also obtain. I remember noting an unusually large in-process inventory in a factory and, inquiring about it, I was told that the employees union encouraged it in the belief that large in-process inventory implied job security!

Usually, though, forensic systems analysts will not be interested in inventory per se, but rather in the effects of poor inventory policy. Nevertheless, the forensic task is to identify root cause of ineffective performance and if inventory policy is established as a root cause, then so be it.

18.3.2.6 Identification and Traceability

Where appropriate, a company will use suitable means to identify products and services throughout operations. What does "appropriate" mean? One answer is that identification is appropriate if it is spelled out in the contract. A general notion is that if a purchased unit used in a final deliverable unit contributes a unique quality characteristic to that unit, then it should be identified.

Parts identification is necessary for three very important reasons. First, the company can verify that the appropriate part is available and being used throughout the production process. Also, the company must be able to identify quality failures by part or lot. The common way to identify a part is through a part number listed on the job order, which then accompanies the part through the production process.

The third reason is that an inventory system based on parts identity and schedule permits evaluation of the dynamics of both logistics and operations. You can index inventory turnover, the first step in improvement. You can determine the cost of storage and in-process inventory. You can track raw materials to machine centers, finished parts to assembly areas; and final product held for delivery. You can determine, and perhaps improve, lead times and queue times. An essential part of inventory and production dynamics is continuous parts identification.

Traceability refers to the paper trail of documents describing all the operations that are performed on a part, including manufacture, rework and inspection, from the beginning to the end of its fabrication. Traceability is critical to accident and failure investigations, as it provides investigating teams with information regarding what was done or not done to a failed part and permits evaluation of such questions as whether a failure occurred due to a processing error, or omission, or a design flaw.

Traceability is the control of an identified product or service and is used if required by contract or by law. Arnold (1994) points out that traceability has two parts, extending rearward to a point before receipt and forward to a point after delivery. The ability to trace a product from its supplier through to the consumer is clearly necessary in the case where it is the active ingredient in a powerful drug. Traceability is often required by government regulation or industry protocol. Even when there is no regulation or protocol, a manufacturer might still want traceability if it envisions having to go back to the supplier, or forward to the customer, *after* delivery of a product. This can come about as a result of customer complaints or of court actions related to liability of product.

A simple example will show the importance of identification and traceability. Suppose that a lot of tuner circuits is installed on a set of 300 television sets of 10,000 sets manufactured in a given period. Later, it is found that this lot was poorly designed. Without being able to identify the lot, it might be necessary to recall all 10,000 sets rather than the suspected 300.

In this example, the traceability of the tuners reduced the magnitude of the problem by a factor of 33. Traceability can have enormous proportions in the automotive industry, where recalls may number in the tens of thousands. Even relatively small manufacturers can limit liability costs if they can trace parts. For example, a manufacturer of mobile shelving systems may rely on several sources for drive motors. If one of the brands develops a systematic fault in the field, then the ability to identify which of several hundred products contain the drive motor can significantly reduce travel as well as replacement costs.

18.4 Supplier Verification: A Case Study

Products are very often made by a team of performers, each sharing responsibility for the overall product. For example, the manufacture of body armor requires a suitable fiber, the weaving of the fiber into fabric, the manufacture of the protective jacket from the fabric, and the testing of the jacket relative to customer requirements. All of these phases of activity may be performed by different companies.

As human life may be at stake in the normal use of body armor, one would expect that the final product must meet quality and reliability requirements, usually mandated by the customer. The metric for reliability of a single use product is the probability of failure on demand. Suppose that the product fails at a frequency greater than that specified by the requirements. Who is responsible in this supply chain? All of the performers? None of them? The final performer? Let us explore this issue with an example.

Company Alpha has developed a very lightweight fiber of remarkable tensile strength that they consider might be useful in making body armor. They contact Company Bravo, a noted vendor of such products. Agreeing on the excellent marketability of lightweight body armor, the two decide to go into the production of body armor classification Type II, which protects against 9 mm full metal jacket and 357 Magnum soft point ammunition. Company Delta is signed on to make the fabric, Bravo will make the jacket and do the testing of the finished product according to Department of Defense (DoD) standards.

18.4.1 Manufacture

Having satisfied itself that its fiber has unusually strong tensile strength, Alpha proceeds with mass production of the fiber without further testing, considering that the quality of the fiber is established. This fiber is shipped to Delta, where it is woven into fabric. The fabric is then tested for resistance to the impact of a steel weight in free fall. This test is considered to establish the quality of the fabric. The tested fabric is then shipped as bolts of cloth to Bravo, which begins manufacture of the body armor. Meanwhile, advertising goes out touting the equality of virtues of the armor relative to others on the market, but weighing considerably less. The newly designed body armor is called *Bravo V*.

18.4.2 V50 Testing

Bravo chooses to follow DoD Military Standard 662F concerning ballistic testing of armor. The test selected is called "V50" and refers to that impact velocity at which half the rounds penetrate and half do not. Rounds of various charge are fired in small groups at samples of the body armor, thus varying the impact velocity. Two factors of interest are recorded: the impact velocity of the bullets and the ballistic resistance of the armor. Relative to the first, the average and the standard deviation of impact velocity are recorded. Relative to the ballistic resistance, again the average and standard deviation of the V50 "go, no-go" velocity are recorded.

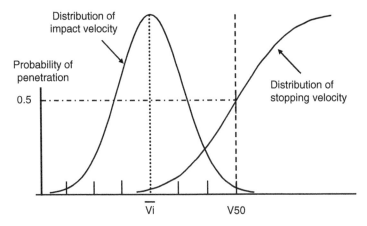

Figure 18.2 The V50 test result of an acceptable design of body armor.

The impact velocity is a function of the range to the target and is unknown, but the muzzle velocity and standard deviation of handguns is known from the records of thousands of firings. For ballistic tests in which the range is very short, which is the case in V50 testing, the muzzle velocity and impact velocity are essentially the same.

By itself, the V50 test establishes only the average velocity of penetration. The quality of the armor is established by how the test is designed. Consider Figure 18.2. The normal probability distribution on the left describes the varying impact velocities during the test firing. The normal cumulative distribution on the right describes the stopping velocity of the armor during the test. In this example, the V50 point occurs at +3σ above the average impact velocity. Therefore, the probability of penetration is less than 0.0014. Most wearers of body armor would be more than happy to wear this armor.

However, the forensic analyst must recognize what has and what has not been demonstrated here. The very low probability of penetration is with respect to the average impact velocity achieved during this test. This average depends upon the amount of powder used in each charge and the spread can be estimated in advance, knowing the characteristics of a given cartridge, say a 9 mm Parabellum. The purpose of the test is to satisfy body armor Type II requirements, so the critical impact velocities are the average muzzle velocities of 9 mm handguns and of 357 Magnum handguns. The 9 mm average muzzle velocity is 1175 feet per second (fps), with a standard deviation of about 35 fps. The 357 Magnum average muzzle velocity is 1400 fps, also with a standard deviation of about 35 fps.

If the actual average impact velocity is substantially below these critical values, then the quality of the body armor is inferior to the requirements and advertisings to the contrary are false claims. Therefore, V50 testing requires honesty and integrity in its design or it can be easily cheated.

18.4.3 V50 Test Results

The *Bravo V* passed all V50 testing and was put into mass production. However, within a relative short time, 24 wearers of *Bravo V* were fired upon with handguns of Type II; seven of them were killed or were seriously injured. Effectively then, the probability of failure on demand was 0.292, or roughly a 30% chance of injury. This was far short of Bravo's claims and resulted in litigation involving all three companies: Alpha, Bravo, and Delta. As all three companies were aware of the intended use of their products, each had a responsibility to ensure the quality and reliability of their contribution. All are responsible for the quality of final product.

There are several forensic engineering questions in determining the distribution of culpability among the companies. Were reliability studies ever made on the fiber, the fabric, or the body armor? If so, where are the data? Were the quality studies of fiber tensile strength and fabric resistance adequate and appropriate? Were the impact velocities in the V50 testing of the body armor approximately those of the critical velocities required for Type II evaluation? These data should all be available in discovery and can help to establish a litigation strategy for false claims.

18.5 Malfeasant Supply Systems

Most of this book is about misfeasant operations—things that have gone wrong because of a nonchalant duty of care, or simply because something was done improperly rather than according to requirements, for whatever reason. Usually, the motivation behind improper performance is unknown or difficult to determine and a complaint focuses on the nonconformance itself.

However, malfeasance also occurs, in which the intent to do the wrong thing is clear to forensic investigation. A few incidents of malfeasance that resulted in the pursuit of fraud and false claims indicate the nature of supplier systems gone very wrong:

- Supplier A knowingly and routinely falsified documents to create the appearance that parts had been manufactured and/or inspected and/or repaired properly, when in fact they had not.
- Supplier B routinely delivered to the prime contractor parts that did not conform to contract requirements, while expressly or impliedly certifying that the parts did conform and knowing that the prime contractor relied absolutely on the supplier's certification.
- Supplier C did not require scan plans of its own vendors, but assigned its own quality assurance agents to falsely record an approved scan plan number on the vendor certification. This was done knowing that the prime contractor did not regularly audit its suppliers. Additionally, the prime contractor

regularly accepted the parts received from its errant supplier and sold them to the customer, aware of the latter's poor quality system and false certification record.

- Company D's supplier system suffered repetitive and sustained failure over the performance period in litigation in numerous areas of contract requirement: supplier evaluation, verification of tools, purchasing procedures, flow down, and internal controls not in place.

References

ANSI/ISO/ASQ (2005). *ANSI/ISO/ASQ Q9000-2005: Quality Management Systems—Fundamentals and Vocabulary.* Milwaukee, WI: American National Standards Institute and the American Society for Quality.

ANSI/ISO/ASQ (2015). *ANSI/ISO/ASQ Q9001-2015: Quality Management Systems—Requirements.* Milwaukee, WI: American National Standards Institute and the American Society for Quality.

Arnold, K. L. (1994). *The Manager's Guide to ISO 9000.* New York: Free Press, p. 115.

Corbett, M. F. (2004). *The Outsourcing Revolution.* Chicago: Dearborn.

Hoyer, R. W. (2001) "Why Quality Gets an 'F.'" *Quality Progress,* October, pp. 32–36.

Lach, A. (Assistant Editor) (2017). "5 Facts About Overseas Outsourcing." *Center for American Progress.* https://www.americanprogress.org/issues/economy/news/2012/07/09/11898/5-facts-about-overseas-outsourcing/. Accessed March 23, 2017.

Pine, M. (2017). "Why Do Companies Outsource?" *U.S. Economy,* February 5. https://www.thebalance.com/why-do-companies-outsource-2553035. Accessed August 24, 2017.

Vitasek, K. (2010). "A New Way to Outsource." *Forbes,* June 1. https://www.forbes.com/2010/06/01/vested-outsourcing-microsoft-intel-leadership-managing-kate-vitasek.html. Accessed September 18, 2017.

19

Discovering System Nonconformity

In the Preface, I quoted the ISO 9000 definition of *conformity* as the fulfillment of a requirement and *nonconformity* as the nonfulfillment of a requirement. While true, this statement does not go far enough. Although a nonconforming unit may be obvious, the task of the forensic systems engineer is to determine whether the process that made or provided the nonconforming unit is itself nonconforming. For purposes of this book, this assessment is called a *finding*. Because the cost of nonconformity can be enormous, all findings will be challenged, so it is essential that a forensic systems analyst have a solid understanding of nonconformities and the kind of evidence required to justify a finding.

Few things are black and white. Some nonconformities are obviously so, but others are not clear. For example, suppose that a company has an apparently high defect rate, but you discover that process capability is not measured. Most standards of performance require a measurement to verify the conformity of units, but each company is permitted the freedom to do so as it chooses. So if

Forensic Systems Engineering: Evaluating Operations by Discovery,
First Edition. William A. Stimson.
© 2018 John Wiley & Sons, Inc. Published 2018 by John Wiley & Sons, Inc.

a company with poor defect rate does not measure capability, is this a nonconformity? It may be, if not under data analysis then perhaps under management responsibility, or quality system requirements, or under requirements for improvement.

The company may assert that improvement is not economically feasible and it chooses to accept the cost of poor quality, which is usually permissible. Many times an analyst must use plain old-fashioned judgment. Some companies could be cavalier about capability or about consumer satisfaction, but if there is a risk of process liability, prudence requires at least some attempt at process control and this increases the variety of evidence that is available in forming judgments.

Sometimes the difference between a major and minor nonconformity is subtle. When an internal control policy or procedure is totally missing, the difference is clear, but when a procedure exists but is not consistently followed, how do you establish that the failure to act is random or systemic? The question is one of common or special cause of failure, and there are few issues of discernment more important to the forensic analyst. In order to determine systemic problems, you look for corroborating evidence of similar type.

Forensic systems analysts must use judgment without being judgmental. This can be done by painstaking attention to detail and collecting of evidence. After you examine the evidence and reach a conclusion, verify it. Remember that in interviewing personnel, there is no record of conversation, so verbal testimony can lead you in the right direction but there should be corroborative records or documentation. Documented evidence provides the objectivity that is the basis of sound judgment. At trial, some of the findings that are based on conversation will be challenged. You cannot count on an interviewed person repeating the testimony in a trial and you will need documented evidence to support the testimony.

For example, suppose that the trail of evidence leads to customer service and the agent describes the routine used in responding to calls from customers. Because of the nature of the job, every call is a potential sale or potential complaint, so that most of what the agent does should have a form associated with it. Ask to see those forms—call chits, order forms, technical service requests, complaint forms—all the paperwork involved with direct customer communication. The agent's job, too, is a process and the record of customer calls describes the task and the agent's performance in doing it.

There should be policies and procedures describing the use of forms for conditional responses. Everything ties together, and you have the task of determining how well this integration is achieved. You are ready for deposition when you have assessed the integration of evidence according to process performance and the bases of findings.

19.1 Identifying Nonconformities

A nonconformity is an event in a operational system that is not in compliance to an applicable standard or in conformance to contract specifications. For purposes of this book in Chapter 14, I defined a *major nonconformity* as one that will affect unit quality or the certification of the QMS. A *moderate nonconformity* may affect unit quality or create problems in QMS certification. Finally, a *minor nonconformity* is deemed to affect neither unit quality nor company certification and would have little effect in litigation.

These are good working definitions in forensic systems engineering. They are reasonable and defendable in court. However, the companies in litigation often have their own definitions and if they differ substantially from yours, then you must be prepared to address this difference. Some companies will classify a major nonconformity as the total absence of a required procedure or a total breakdown of a procedure or process. A minor nonconformity may be a single observed lapse of a procedure or process failure, which in the estimation of the observer, is not a systemic malfunction. However, if the forensic analyst estimates that a single observed event may lead to systemic failure, then the nonconformity may be reclassified as major. Some companies call a major nonconformity a finding, and a minor one an observation. Others use *observation* to describe an existing or potential nonconformity.

Dennis Arter (2008. E-mail letter of July 3, 2008. Response to inquiry on behalf of the American Society for Quality), Fellow of the American Society for Quality, reports that these local usages have caused widespread confusion as they depart from the original idea that an observation was simply that, and carried no connotation of conformity—it was neutral. As observing is fundamental to forensic investigation, this book uses *observation* in its original meaning: a statement of fact.

You cannot ignore the parochial definitions used by the opposing side because they will use their own definitions in court. If using the same terms, their definitions conflict with yours, the jury will become confused and once again, the outcome will be based on the persuasion of counsel rather than on the evidence.

Perhaps the best resolution to this confusion is to drop the classification of major and minor altogether. The terms are not used in ISO 9000 and have no universally accepted definition. Duke Okes (2008), also a Fellow of the American Society for Quality, notes that these terms "do not provide sufficient differentiation as to the level of action required." Okes suggests the most effective way to classify nonconformities is according to their frequency and risk. Simple metrics could be used, such as "low, medium, or high." Priority follows automatically. High frequency and high risk would have top priority; conversely, problems of low frequency and low risk might be deemed as minor.

This strategy is used in Chapter 14 and is recommended as the optimum approach.

19.1.1 Reporting Nonconformities

Nonconformity is a serious business in the evaluation of operations and deserves special attention. The forensic systems analyst may take notes on various areas in the investigation, but the observation of nonconformity must sooner or later be formal because it will be integrated into the initial or amended complaint. Generally, the analyst will work with counsel in choosing the most effective phrasing.

Technical people often ignore the importance of words, counting on mathematical expressions to firm up their arguments. But at trial, words are as important as evidence. When several people, however honest, testify about the same event, the Rashomon Effect ensures that you will get several different stories (Kurasawa, 1950). The evidence is often outweighed by the telling of it.

Or consider the US Constitution, written in English by educated men whose native English language had been in their families for centuries. Few documents have been better written. And yet, it can hardly stand as undeniable evidence. Wise men and women have been debating what it says for over 200 years.

In particular, forensic systems engineers should be aware that their terms of art may be misunderstood, or that counsel might prefer some other term. For example in referring to a process, engineers may say "out of control" rather than the longer "out of statistical control" because other engineers will know what is being said. However, "out of control" has a stark meaning to the general public and will be vigorously resented by the defense. It is counsel who is responsible at trial. Therefore, it is counsel, who is responsible for the litigation strategy, the tactics to be used, and the language to be used.

So you begin by noting the workstation at which the observation was made, and its location. If the process or machine has a name, write it down. You will want the names of those persons directly or indirectly associated with the nonconformity because such persons may be required in deposition or at trial. It is very important to describe accurately the location of the nonconformity and the time and duration of its existence as this helps to identify the problem, distinguish it from other similar nonconformities, and to determine whether the event is random or systemic.

Write down the applicable requirement: the one to which the event is nonconforming. If there is more than one, write them all down, pointing out how the event is nonconforming to each. If you are unsure of whether several requirements are applicable, write down the one that you are sure about. If you are unsure about exactly which requirement is applicable, be prepared for a fight because the report will surely be challenged. The rationale for listing several requirements, if they are all applicable, is to demonstrate the breadth of impact that the nonconformity has on operations, thus emphasizing its severity relative to risk.

An example of applying several requirements to a nonconformance would be a workstation at which there is no procedure for what to do with nonconforming product. Such product can be reworked or scrapped on the operator's own initiative, but in any case it must be removed from the production line.

If the forensic analyst is investigating delivery of nonconforming product and traces the problem to this workstation, then possible requirements that apply are those referring to the provision of parts; those referring to measurement and monitoring; and those that refer to the control of nonconforming product. The only way to be sure about all of these is to track down the evidence. You must consider all possible applicable requirements. If you find evidence with respect to only one requirement, then write it down as the applicable one.

Describe the nonconformance as accurately as possible. If there are several ways to describe it, use words to describe the most serious condition. For example, if a pool of oil is on the floor in an area where operators carry material, don't just say that you observed a pool of oil but include the safety risks in the description. If a document revision is out of date, a properly worded observation will describe the effect of this on operations. It might also be worthwhile to consider what flaw in the process allowed a document to become outdated without automatically issuing a current one. Then this nonconformity may serve to be the first of a string of other similar omissions, indicating systemic process failure.

19.1.2 Disputes

The observations of a forensic systems analyst concerning process operations are those of an outsider. In a litigious environment, they will almost always be challenged and the nonconformity disputed. If the defect is estimated as high risk, there may be a great deal at stake. Therefore, the analyst should take care to use correct wording in the assessment. Despite having a strong evidential nature, a badly worded report can be a disaster.

For example, suppose that you are visiting a manufacturer of plastic film whose operation is in litigation because of a systemic production of nonconforming product. An important process in this product is the calender, a machine consisting of a series of hard pressure rollers used to form or smooth a sheet of material. During the visit, you notice a row of five calender machines and in conversation with an operator, you find that one of the machines was malfunctioning during the period in litigation. This machine could be a major contributor to the problem, but if you inadvertently cite the wrong one, your case can be seriously damaged. Your error will not be considered a "typo"; rather, your competence will be challenged, the wronged machine offered as evidence of your error, and the issue of the actual nonconforming machine will be finessed.

In 1996, a *ValuJet* aircraft crashed in the Florida Everglades, killing 105 passengers and 5 crewmembers (National Transportation Safety Board, 1997). Investigators found that the crash was caused by a cargo of full oxygen generators that had been loaded aboard without their safety caps. Without the caps the generators activated, producing intense heat, which subsequently caught the aircraft afire. Mechanics working for an airline maintenance company were supposed to follow policy and procedures for the transport of oxygen generators. The policy was that the generators must have safety caps if they were not empty. The procedure called for labeling of empty generators. During a long period while criminal charges were pending, no one would admit to anything, fearing that anything that they said would be held against them. Certainly, in litigation much less dramatic than this, forensic investigators will run into stonewalling from time to time. The best approach to take is to gather credible evidence and to be correct and specific in describing the observation and its characteristics.

19.2 The Elements of Assessment

Typically, litigation involving business operations involves the delivery of one or more nonconforming units in which a plaintiff suffers physical or monetary injury. The existence of a nonconforming unit is essential to this process and an expert witness may be called in to establish the nature of the nonconformity. For example, a professional engineer may be called upon to conduct fractography of a metal or plastic unit whose failure is believed to be crucial to the complaint. The subsequent report will describe the technical characteristics of the unit in terms of its defects, if any.

The forensic systems analyst is concerned with the possibility that the process of creating the unit may itself be nonconforming, suggesting a causal influence and a larger scale of nonconforming product. The subsequent process report will describe the technical characteristics of the process in terms of a *finding*. The wording of a finding is critical and we shall examine this issue later in this chapter.

19.2.1 Measures of Performance

It may well be that several processes of operations are involved in litigation, perhaps all of them. This calls for a top-down assessment of conformance of operations and an assessment of each pertinent system in terms of their effectiveness and efficiency. Recall from the Preface that effectiveness is a measure of the extent to which planned activities are realized and planned results achieved. Efficiency is a measure of the relation between the results achieved and resources used. The degree to which the planned activities, results, and used resources all meet or exceed contract requirements indicates how well overall operations are performing. Therefore, effectiveness and efficiency are necessary and sufficient measures of a corporate quality management system (QMS). Various metrics to measure the effectiveness and efficiency of processes were discussed in Chapters 11, 16, and 17.

19.2.2 Considerations in Forensic Analysis of Systems

Forensic systems analysts will review the evidence in discovery relative to the processes of interest, framing their observations in the context of six questions:

1) How does this process work?
2) Is the process in compliance?
3) Is the process performance acceptable?
4) What are the key measures of how well it works?
5) Is there a nonconformity?
6) Is the nonconformity causal?

The compliance of the process to requirements is a straightforward evaluation of verifying and validating system *form* against that of the requirements. Noncompliance may be a breach of contract in itself, but if there is no adverse impact to the process output, the issue may not be worth pursuing. The next four questions address the assessment of performance and require an evaluation of the *substance* of the system, the determination of appropriate metrics, possible statistical measures, and the analysis of results. The last question addresses process liability and requires inductive and deductive reasoning in forming findings. If the nonconformity is random, it may not be worth pursuing. If the nonconformity is causal, then the possibility of systemic failure should be considered and a further search for evidence should be sought.

Counsel may suggest your path of investigation or leave the matter to you. I prefer to begin by reading the complaint. This is usually a long document, but it outlines the charges and areas of most probable cause for process liability. Then I begin a strategy of research that is based on the estimated risk factors suggested in the complaint. One way to categorize the risk factors is by corporate function, for example management, documentation, supplier control, and verification and validation.

Counsel may also suggest what evidence should be considered first. If there is no pressing matter, you can choose any order that seems reasonable to you, so long as your investigation is comprehensive. Evidence that is neutral in terms of liability is put aside. Evidence of nonconformity followed by corrective action may also be put aside, depending on its duration. Some corrective action takes a considerable period of time. Some corrective action is moribund and may be considered nonconforming in itself. Evidence that suggests nonconformity is then filed in an appropriate category and retained for later consideration relative to process liability.

19.3 Forming Decisions

Forensic systems engineering is a decision-making process. You study the evidence in discovery with the sole objective to come to some kind of conclusion on whether the controls of operations are effective or not. This study is conducted process by process and in doing so, you use certain terms to describe what you see and your conclusions about the effectiveness of the controls. There are three terms that are extremely important for analysts to understand. The integrity and utility of the assessment depend upon them. The terms are described in Table 19.1 and discussed in the following paragraphs.

Objective evidence is the data and information the analyst gathers in the performance of the investigation. Formally, the term is defined as "verifiable qualitative or quantitative observations, information, records or statements of fact" (Russell, 1997). The evidence must be verifiable. Examples of objective evidence are measurements, records, and testimony, if the latter is verifiable. Personal statements can be considered as objective evidence, but only if they are verifiable. Hence, rumor and hearsay are excluded. You can use word of mouth as a basis to hunt down verifiable information, but you often cannot trust the statements themselves. Some employees may boast of their performance to promote their own position, or they may bad-mouth their employer for some real or imagined transgression that has nothing to do with the litigation. Qui Tam attorneys are experienced in distinguishing between disgruntled employees and true whistleblowers, but even an honest witness can unknowingly be subjective in testimony.

An *observation* is sometimes defined as the objective evidence of nonconformance. In other words, an observation is a deficiency. In my view, this is not

Table 19.1 Analyzing observations to form decisions.

Objective evidence	Verifiable information or data
	For example, measurements, records, testimony
Observation	An item of objective evidence
	For example, analyst's notes derived from evidence
Finding	A conclusion of importance based on observations
	For example, document control system not effective

a good definition because it leaves undefined the evidence that *verifies* a controlled activity. A much better definition is given by Russell (1997): "An observation is an item of objective evidence found during an audit."

A *finding*, on the other hand, is the conclusion of the analyst about a control, a process, or an entire system and is usually taken to mean a statement of nonconformity of high risk supported by observations. It is the logical and concise formulation of all observations summarized to a probable cause. As an example, grease on the floor is not a finding. "Unsafe working conditions" is a finding. Similarly, an out-of-date document at a workstation and an unsigned job order are observations leading to a possible finding of "Document control system not effective."

Table 19.2 provides additional clarity on the differences between an observation and a finding. Notice in the table that the absence of sign-off activity was endemic—found in three different departments. Each is an observation, perhaps made by different auditors. Yet, there is only one finding. Findings are the very important summary conclusions that concisely describe the effectiveness of a process. Counsel will focus on them as leads to litigation strategy and will choose the wording. I have witnessed a number of different wordings for a finding of, say, a production system. "The production system is out of control, nonconforming, inadequate, ineffective, improper in systemic failure."

Usually, there will not be too many findings even if there are many observations. Too many findings could overwhelm a jury and cause confusion about exactly what the grievance is. As a simple example, if the analyst determines that, say, there is large-scale escape of nonconforming units, why confuse the jury with the 2000 documented escapes? The latter would be critical to the case and provide supporting evidence to the finding, but no jury wants to read

Table 19.2 The nature of a finding.

Finding

Document control procedures are ineffective. Specifically, the document authorization procedures do not support corporate policy

Observation no. 1

Design change *Eng06: Engineering Change Order no. 1522*, dated October 18, did not have an authorizing signature on the signature sign-off plate

Observation no. 2

Job order no. 2618, dated November 30, located at the shipping dock, was not signed off by the production manager per procedure *Prod 222*

Observation no. 3

Purchase Order no. 095, issued to vender Bravo on October 27, did not have an authorizing signature on the signature sign-off plate per procedure *Pur15*

2000 documents. ("Escape" is used in some industries to describe the unintended delivery to customer of a nonconforming unit.)

19.4 Describing Nonconformities

An assessment of performance answers six questions: what, why, who, when, where, and how? The descriptions of nonconformities shown in Table 19.3 may be critical to the successful pursuit of justice. The best way to describe a nonconformity is systematically, which is why the six-question approach is effective. When you have answered the six questions, you have described the nonconformity so that it can be correctly identified, its impact on performance understood, and the case for liability established.

What refers to the nature of the nonconformity. This description goes beyond simply stating that the nonconformity is high risk. The following are some examples. There are no documented procedures for a process. There is no corrective action on nonconformity. There is no calibration program. There is no approval process for changes in drawings or specifications. There is no review process on engineering changes. A policy exists concerning evaluation of subcontractors, but there is no procedure. Policy and procedures exist for subcontractor evaluation, but there are no records.

Where refers to the specific workstation, machine, or process where the nonconformity was located. It is important to get this right, first because it indicates the owner of the problem, and second because it helps counsel to locate the matter at some later time. An error in the report can lead to a dispute, as shown earlier in the example of a malfunctioning calender, one of five, the others being good. As another example, an activity may have three identical stamping machines. One of them has an out-of-date work instruction. Suppose that you inadvertently record the wrong machine number. Later, defendant management finds a current work instruction there. Your entire report could henceforth be suspected.

Table 19.3 The elements of nonconformity description.

What	Nature of the nonconformity
Where	Process nonconforming
How	Manifestation of the nonconformity
When	Date and time or duration of the nonconformity
Who	Responsible performer or process owner
Why	Apparent cause of the nonconformity

How means how the nonconformity is manifested—its physical appearance. Often this is the same answer as "what," but not always. For example, suppose that a large set of steel cabinets are delivered to a customer, many of them with bent sides. A dispute arises as to culpability: the manufacturer, the delivery system, or the customer receiving station. In your investigation, you learn that at the manufacturing plant a forklift operator had laid a layer of steel sheets on a small box. The sheets, intended as siding, were seriously bent but were integrated into production anyway. Therefore, the "what" here is a nonconformance to the proper handling requirements and the delivery of nonconforming units, but the "how" is the defective placing of the sheet steel such that it became permanently bent. The act described by "how" not only helps to locate the nonconformity, but it also describes the immediate conditions surrounding the event and may indicate the cause.

When, in forensic work, refers to the date and time of the nonconforming event. You should also record the date and time of any corrective action if such were taken. The time of the event is a very necessary index for establishing systemic failure, either because such nonconforming events were frequent and serial, or because one nonconformity remained unresolved over a sustained period.

Who refers to the person or persons associated with the nonconformity. If you name any persons, the role of each person in the nonconformity should be identified. This is not done to attach blame—which problem will be addressed by counsel—but serves to identify any decision makers related to the dynamics of the event. For example, if improper procedures were used, then it is important to establish whether the person(s) doing so were competent and authorized agents. Depending on this identity, a finding could range from an ineffective training program to malfeasance and fraud.

Some managers are opposed to interviewing operators during an investigation of operations on the grounds that the performance of operations is management purview. The antecedent of this argument is correct, but the argument itself is disingenuous and effectively precludes important testimony. Usually, counsel will suggest the persons to be interviewed by the forensics analyst, if such interviews are deemed necessary to the litigation. The reader will recall from the Preface that interrogatories, admissions, and depositions are admissible devices of discovery.

Why refers to the apparent cause of the nonconformity. Although identifying the cause is necessary to effective correction, by the time the forensic systems analyst is called in, correction is a moot point. The analyst must identify the cause of nonconformity because as discussed in Chapter 15, the cause can be crucial in establishing process liability. Just as the existence of a body is required to pursue a homicide, and a motivation is required to pursue a crime, the successful pursuit of process liability requires a cause of nonconformity.

From a systems engineering point of view, in the event of unit nonconformity, there are two causes to be identified: (i) the process that caused the unit nonconformity and (ii) the cause of process nonconformity. The first must be established because often (and perhaps usually) the law does not assume a causal relation between a process and its units. It must be proven. The second cause must be identified because the assignment of nonconformity to any process will almost always be disputed. It will be disputed because whether or not the law agrees, every manager knows what every engineer knows: that there is a cause–effect relationship between process and product and the defense will fight an assignment of process nonconformity vigorously.

19.5   A Forensic View of Documented Information

Invariably, the evidence of nonconformity will be in the form of documented information. By that is meant the recording of information regarding an event or a process. It is important to understand that the critical issue is the message and not the medium. Hence, documented information refers not just to paper media but electronic, video, and sound—any medium that can preserve human expression. If the evidence of interest is a testimony, it will be corroborated by documented information, hereafter called "documentation" for brevity.

Documentation itself can be a high-risk issue in the litigation of operations. Indeed, the issue of documented evidence was listed in earlier chapters as one of the functions of operations that appear most troublesome in litigation, along with process management, supplier control, and verification and validation. The forensic approach to verification and validation and to supplier control are discussed in detail in Chapters 16 and 18, respectively. The approach to process control is discussed throughout the book. Now let us tackle the problems of documentation.

In recent years, there has an energetic trend to a paperless society and is generally achieved by simply transferring information from paper to computer. There is also an effort to reduce the amount of documentation required to conduct business because maintaining electronic documentation may lighten the spatial burden but not the volume. And as the Internet has clearly demonstrated, you can have information overload in the cloud also. Moreover, an electronic system presents new obstacles to security, fraud, and durability. None of this should discourage the forensic systems analyst. Tracking down information is a hard work, no matter the medium.

Documentation control is often approached with some reluctance by investigators. First because the effect of poor document control on operations is often underestimated and secondly, it is difficult to find a subject more voluminous and tedious to study. A litigation of contract performance can easily involve hundreds, if not thousands, of documents: manuals, drawings, specifications, orders, schematics, standards, notes, emails, records, reports, contracts, and more. After perusing this stuff for several hours at a time, your eyes tend to glaze over or you tend to nod off. And yet, it is in the documentation concerning an event that we most often find the truth about it. Documentation provides solid evidence. So let us bite the bullet and push on.

19.5.1 Requirements in Documented Information

In litigation of corporate operations, the objective of the forensic systems engineer is to examine the performance of the operational processes in order to verify their compliance to form and their conformance to substance in accord with the contract. The analysis begins with a top-down approach to understand the organization of operations and to identify all the processes therein. The organization's QMS indicates the trail and relationship of documentation requirements. The first document—the first among peers—is the company's quality manual.

19.5.2 The Quality Manual

If a performer's contract calls for a formal QMS, then to meet this requirement the performer will document the system. This document is usually referred to as a QMS manual or simply a quality manual. In this book, because of its international reputation and for demonstrative purposes, the quality manual will be based on ISO 9001.

The quality manual is of primary importance to the forensic systems analyst because, having won the contract, the performer will argue that the company's quality manual is the guiding document and not any specific standard. There should be no difference between the two, especially in the case of ISO 9001, in which auditors will have reviewed the quality manual for compliance before certifying the company. Nevertheless, there may be differences. Some standard requirements may have been waived by the certifying body or by a given contract, in which case the performer owes no duty to them.

In the general scheme of ISO 9001, a QMS consists of two parts: the documented part and the implemented part. The documented part is the quality manual and includes or refers to all those documents: policies, plans, procedures, instructions, and records that affect or refer to the quality of operations. The manual need not be a formal volume or set of volumes, but it can be an assemblage of printed matter and electronically recorded programs. However, the totality of documents must be defined and controlled to ensure the availability of current documents where and when needed and to keep the total documentation from getting out of hand.

The entire assembly of documents can be considered the quality manual—indeed, ISO 9000 refers to the quality manual as all documentation used to

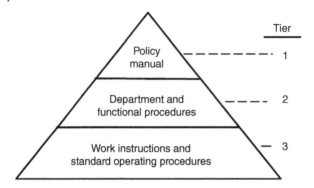

Figure 19.1 Quality documentation arrayed in tiers.

specify the quality system. But many companies assemble a variety of documents and software as their total quality documentation and present a small policy document that they call the "Quality Manual." This arrangement is shown in Figure 19.1 and is justified as a quality manual if it formally refers to the totality of quality system documentation.

The Tier 1 policy manual contains policies and responsibilities related to the governing requirements. It expresses a written commitment of the top management to the quality of its products. It also contains the list of Tier 2 quality documents with associated revision lists and defines the distribution list for the manual. The manual may also contain corporate-level procedures, such as steering committee procedures. It may also outline the entire QMS and in this case it is sometimes called the quality manual, although strictly speaking the quality manual is the entire body of quality documentation.

Departmental procedures explain the responsibilities and provide the organization chart of all departments or activities, and describe the input and output interfaces of its processes. Also, just as the quality manual provided an index of Tier 2 documentation, so also each department procedures provide an index of the department's Tier 3 procedures.

A work instruction, sometimes called a standard operating procedure, is a document that describes, step by step, how a task is to be done. Because there is sometimes confusion on this matter, a work instruction should be distinguished from a *job description*, which is a document that describes what tasks a given employee performs. A job description describes *what* an employee does. The work instruction describes how to do the task. It is located on-site and tells how to set up the machine or equipment with respect to a given job order.

For example, a work instruction for a lathe does not describe how to do machining; it describes how to set the machine up. Similarly, running a test on a centrifugal chiller requires highly skilled electronic technicians. Despite their skill, the tests also require work instructions that describe test station set up for the particular kind of chiller to be tested: loop temperature ranges, pump pressure ranges, and so on. The purpose of the work instruction is to establish the conformance between the range of operation of the machine or equipment and the specifications on the job order.

Not every task requires a work instruction. You can determine whether or not a task requires a work instruction by asking two questions: (i) If the task is done incorrectly, will it affect the quality of the unit? (ii) Is the task done self-evident to a new employee? Yes to the first question means that a work instruction is required. Yes to the second question means that a work instruction is probably not needed, except possibly to provide data. For example, engineers do not need work instructions to design things. The task is self-evident based upon their qualifications.

As a forensic systems analyst, your interest regarding work instructions is primarily in whether the instructions are at issue in a litigated process. Mainly, your focus is on whether the litigant's quality manual and the system in operation are in agreement. Assuming that the quality manual meets contract requirements, then it says what the performer should be doing. The actual activity of operations describes what it is really doing.

19.5.3 Documented Information Control

Good business practices include control of the distribution of all documents and data related to the quality of product and service, with associated procedures that describe the method of control. Such documents attest to the quality level of the unit and to process performance and include specifications, procedures, drawings, blueprints, and many types of records.

The purpose of document control is to ensure that current and correct documents are available where and when needed, and that they are used. From this definition, we can construct the necessary controls:

- Identification
- Collection
- Indexing
- Disposition
- Amendment

Identification refers to the document title and revision date. Every quality document is assigned an "owner" who is responsible for its control. The owner and users identify each document by a name and control number. Documents and data may be in hard copy form or stored in a computer.

Collection means that there is an orderly and accountable distribution of documents. Only controlled copies are to be used and copies must be easily available to users. The forensic analyst may find spurious or off-the-cuff documents having been used, perhaps created as a jury rig to solve a current problem. Processes and units using such documents may well be nonconforming and will require further investigation as to their validity.

Indexing means that there must be a master list of current documents, each document having a current date, revision number, and approval signature to ensure document validity. As we described earlier, the master list of Tier 2 documents resides in the policy or quality manual. The master list of Tier 3 documents resides in the Tier 2 document associated with them. For example, the index of work instructions within an activity will reside in the activity procedures book.

Disposition is the issue of new or revised documents and removal of obsolete documents. There should be a policy and procedure for the issue of new or amended documents and the removal of obsolete ones. Only currently authorized documents should be in place, according to appropriate policies and procedures.

Amendment refers to making changes to controlled documents, for which there must be a procedure. This procedure should describe the following: (i) How changes are identified within the document; (ii) How and when current revisions are identified; (iii) The number of revisions before a document is reissued. Some documents and forms such as test reports and purchase orders seldom change, and there are those that often change and so have revision levels—for example, drawings, procedures, and specifications. The process for revision is to (i) identify the level on the document; (ii) review and approve changes to documents at the same level; (iii) identify the nature of the change; (iv) maintain a master list; and (v) reissue the document after a practical number of revisions.

Examples of documents that require control are drawings, specifications, blueprints, test procedures, inspection instructions, work instructions, operation forms (when filled out), quality manual, operational procedures, and quality assurance procedures. Each company can choose which documents it wishes to control, but there must be a policy and a procedure to describe it. In some cases, customer requirements dictate the use of certain documents such as technical manuals associated with specific equipment. In my experience, performers are usually quite good at this, but the currency and relevance of documents related to the goodness of units must be verified in forensic investigation.

19.5.4 Records

A record is a document that provides objective evidence of activities performed or results achieved, and that is retained. The retention is why it is called a record. The evidence therein may include statements, graphics, images, voice, and any other medium that can be recorded. Quality records are those that demonstrate conformance to the specifications, effectiveness of the quality system, and conformity of system outputs. Test records are especially important in forensic investigation as they testify as to the status and conformity of the units tested and may be a critical issue in litigation.

How do you distinguish between documents and records? Eugenia Brumm (1995) answers the question simply: A document exists before the fact; a record exists after the fact. Thus, a document is a graphical or written description of a policy, procedure, or instruction regarding an activity. A record, on the other hand, contains data, information and perhaps a conclusion resulting from that activity.

The statements on record are presumed to be true by the author of the report. At this point, it should be understood that in logic, a statement is defined as a sentence that is either true or false (Manicas & Kruger, 1976). If it is believed to be true, then it is called a statement of fact. An argument is a set of statements, and a conclusion, or judgment, is an opinion derived as a consequence of the statements.

Nevertheless, not all the statements in a record are true. On rare occasion, a forensic analyst will find false statements in a record or a report. More often, the statements may be true but the activity itself is false. For example, suppose that a certain test specification requires x-ray films taken of a unit from four different angles. Each film requires a 5-minute set up time and there are hundreds of units to be filmed. The production manager, seeing a serious delay in throughput, reduces the filming to two different angles. Suppose further that all the films come out in conformity and this statement goes into the report: "All units tested in conformity." The statement is true; the test is not.

Retaining documents can be a voluminous business. Therefore, most management standards specify a set of records that must be retained. The company is then free to choose which records to keep beyond those specified. There may be good reasons to do so beyond the concerns of product conformity. I once visited a fabrics plant that retained thousands of pieces of cloth of different material and color. It turns out that a certain color, say cobalt blue, on one material will have a different hue on another. To reduce the cost of experiment, the company simply saves a sample of any product that it makes, thereby quickly meeting similar future customer requirements up front.

In the pursuit of evidence concerning operations in litigation, typically, the forensic systems analyst will be concerned with records of the following:

- Evidence of conformity to requirements and of effective operation of the QMS
- Management reviews
- Personnel education, training, skills, and experience
- Conformity of operational processes and resulting product or service
- Design and development reviews, verification and validation, and change orders

- Supplier control
- Validation of processes for operations
- Customer property
- Control of monitoring and measuring equipment
- Internal audit results
- Monitoring and measurement of product
- Control of nonconforming product
- Corrective action

Knowing the nature and purpose of the various documentation associated with a given litigation, the forensic systems analyst can more easily identify potential evidence. Then understanding how documentation systems are properly maintained, the analyst can more quickly determine the validity of a given document as evidence, and just as important, whether the document is nonconforming itself.

Acknowledgment

Some parts of this chapter concerning documented information were adapted from the book, *Internal Quality Auditing* (2010), © William A. Stimson. Paton Press, Chico, CA.

References

Brumm, E. K. (1995). "Managing Records for ISO 9000 Compliance." *Quality Progress*, January, pp. 73–77.

Kurasawa, A. (1950). *Rashomon*, a film whose plot concerns various characters that provide contradictory versions of the same incident. The film revealed a truth about human nature that has become so universally recognized that the human characteristic is now called the Rashomon Effect. Japan: Masaichi Nagata and Minoru Jingo.

Manicas, P. T. and Kruger, R. N. (1976). *Logic: The Essentials*. New York: McGraw-Hill, p. 41.

National Transportation Safety Board (1997). *In-Flight Fire and Impact with Terrain Valujet Airlines Flight 592DC-9-32, N904VJ*. https://www.ntsb.gov/investigations/AccidentReports/Pages/AAR9706.aspx. Accessed September 18, 2017.

Okes, D. (2008). "Are Your Audit Nonconformances Nuisances or Problem Statements?" *The Auditor*, November–December, pp. 7, 13.

Russell, J. P. (1997). *The Quality Audit Handbook*. Milwaukee, WI: ASQ Quality Press, p. 203.

Appendix A

The Engineering Design Process

A Descriptive View

When we think of *design*, great artists and builders come to mind, such as Michelangelo, Leonardo da Vinci, Berthe Morisot, or Christopher Wren. We imagine that design is mostly creativity and certainly it is that, but there is much more to it. True, the essence of design is creativity. No standard in the world can mandate or bestow creativity. You either have it or you don't, but creativity is brought to life through a design process that is well identified. For example, Michelangelo followed a detailed process in both sculpting and painting, having to make his own pigments from scratch; having to design his own scaffolding; and having to put together a competent team (King, 2003). As another example, in engineering, textbooks such as that of Arora (1989) describe a formal and mathematical approach to design.

The Six Sigma philosophy calls the process of design "Design for Six Sigma" and defines it roughly in these steps: define, measure, analyze, detail, and verify (Six Sigma, 1998). In similar terms, the standard ISO 9001 (2015), too, offers a sequence of detailed requirements to accomplish design: from the planning phase through the phases of inputs, outputs, controls, reviews, verification and validation, and on to the change processes inherent to design and development.

Forensic Systems Engineering: Evaluating Operations by Discovery,
First Edition. William A. Stimson.
© 2018 John Wiley & Sons, Inc. Published 2018 by John Wiley & Sons, Inc.

Every company designs something, if only in personalizing some function—say customer service. In a very real sense, design and planning overlap, so that any planning function designs and any designing function plans. Design is a dynamic activity and proceeds tentatively toward a goal acceptable to the designer. Hence, the process is best described as "design and development" and if properly organized, the result is sustainable. A sustainable design is one in which the creativity won't get lost or inadvertently abridged.

A.1   Design and Development

Design and development refers to the planning, inputs and outputs, amendments, improvements, verification, and validation of the design of products, services, and processes. The design process includes life cycle, ergonomics, reliability, maintainability and sustainability issues, as well as the risk issues of safety, disposal, and the environment.

A.1.1   The Design Process

The design process tends to be similar in every industry—software, hardware, product design, system design—all use a process as shown in Figure A.1, which, with interconnecting feedback loops in every phase of activity, allows continual opportunity for reiteration and development. In some specific litigation in which a particular set of design procedures are identified, the forensic systems analyst may audit a questionable process with respect to them. However, the processes shown in Figure A.1 describe the general procedure.

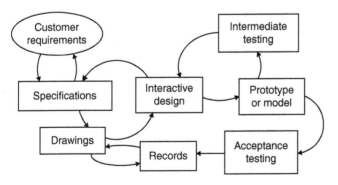

Figure A.1 The design and development process.

A.1.2 Customer Requirements

In the industrial world, the basis of a design begins with a set of customer requirements that are transformed into specifications. This is more than just coming up with numbers; it also means determining materials, drawings, skills, and all the resources required to achieve this transformation. ISO 9001 requires that this process be formalized. A performer is free to choose its design process and there are many models to choose from. Nevertheless, most design processes have similar characteristics as shown in Figure A.1, which depicts a model of closed-loop mechanisms that are required for robustness and that make the design process effective (Stimson, 2001).

The process may include development of an initial design into various prototypes or configurations that will be preliminary to a final design. The specifications are transformed to drawings and go through series of cycles of creation, redesign and intermediate testing, first on paper with various mathematical models and later with more costly prototypes.

A.1.3 Interactive Design

The phases of development may call for in-process testing; this is particularly true in the design of software programs or complex hardware systems. Figure A.1 depicts the feed forward of requirements to test program formulation, as well as feedback of results following some initial progress. There is a feed forward to configuration control, which management is an important part of the process. In the early stages of design in particular, when requirements may be fuzzy, configuration changes can be undisciplined. Without a paper trail it may be difficult to review why design changes were made. This loss of information can limit design reviews and the reevaluation of alternative designs.

Development, test, and configuration require constant iteration in this closed-loop cycle because it is in the iteration that a final, acceptable product is possible. Although there may be time constraints in design, perhaps because of contract requirements, design is a life-cycle process. There is always a feedback loop in permanent structure so long as a given product is on the market because no matter how well the final design fits requirements in the lab, the in-service life of the design is the acid test.

A.1.4 Intermediate Testing

In-process testing is often used in the development cycle of software programs or complex hardware systems. For example, an intermediate testing phase feeds back into an iterative design scheme, because as you develop a design it may become apparent that you "can't get there from here," or that a different approach offers improvement.

Indeed, it may even be necessary to change the specifications. As this is tantamount to changing the customer requirements, the customer cannot be left out of this loop. Indeed, the design process must allow for adaptability at any stage. This flexibility imposes two obligations on the design team. First, all interested persons, including the customer, should participate appropriately in any design amendment considerations. Second, configuration management must be maintained. As pointed out earlier, in the early stages of design when requirements are translated to specifications, changes can be undisciplined and evaluations left unrecorded.

The purpose of the intermediate testing is to verify and validate the design. It cannot be repeated too often that verification and validation are two different entities. Verification is the act of ensuring that the design meets the specifications. This is achieved through holding design reviews, conducting feasibility tests and demonstrations, determining alternative calculations, and very often, comparing the new design with a similar proven design, if such is available. This step is a type of benchmarking and lends credibility to the design.

However, benchmarking can also be misleading, both for verification and for validation, if the specifications and intended use of the new and old are not similar. This warning is particularly appropriate in reliability design, where errors in judgment are likely if the comparison of previous and new designs is based on an insufficient number of data.

Validation is the act of demonstrating that the product or service meets the customer's intended use requirements. Validation must follow verification because the performer must first prove that the required unit under test has been built. It is meaningless to validate a unit that has not been verified as meeting specifications.

A rule of thumb for the design of a relatively simple product is that validation takes place twice, once at agreement on contract and once following acceptance testing—that is, before delivery. This rule is completely inadequate for the design of many products and complex systems. In validation, both a contract review and a design review are appropriate for each development cycle. It is simply too easy to lose sight of the goal in a process as introverted and tunnel-viewed as design. John Gibson (1990) often referred to the need of software people to just "jot down a few lines of code to get things started." This temptation almost guarantees a divergence from the design as it is and the design as it should be.

The design process requires horizontal and vertical interfaces. *Vertical* interfaces are those communications in a line organization related to a particular job order. *Horizontal* interfaces refer to the exchange of ideas, data, and information between appropriate groups while the design is in a particular phase. For example, early design meetings should include marketing, customer service, production, purchasing, and the customer. Purchasing people provide a reality check for materials costs and availability. Production people ensure the agreement of requirements to performer capability.

The design output must be documented, verified, and validated with respect to the design input requirements in each phase of the development cycle. Crucial characteristics of the design must also be identified at this point, such as safety, storage, weight, and other features required by law or critical to operation. Final design is often preceded by iteration of preliminaries. The design team can be flexible with intermediate designs, but final designs must be documented. They must also be reviewed for conformance, verification, and validation prior to release. This confirmation usually takes the form of well-designed acceptance criteria, which is as important as the product design itself.

Review meetings should be formal, documented, and include all the players concerned with the design. The outcome of these design reviews may be considered a legal document that should be retained in the event of future litigation. The scope of the design has a great deal to do with the magnitude of the review. A large software program, for example, may take over a year to write, and the design process would include many reviews and walks-through. Contract requirements, such as ISO 9001, will be concerned that the design process be formal, suitable, and documented.

Amendment procedures must be formal. This is a universal step in the design process and makes sense with or without ISO 9000 considerations. The procedures should include documentation, review, and an approval process before a change can be implemented.

A.1.5 Final Iteration

ISO 9001 requires that all the design subprocesses be formalized, not necessarily in organizational structure as in documented operations. Design, development, and planning responsibilities and interfaces must be identified, as well as customer input. The paper trail from customer through acceptance testing must be recoverable for those documents that pertain to product verification and validation.

The paper trail of design is critical for two reasons: one for design purposes and one for legal purposes. It is not necessary to retain brainstorming or intermediate notions about a product's design; documenting fuzzy ideas would inhibit the creative process. But as the design is developed, its connection to the customer requirements must be traceable. This is because traceability permits later reappraisal of earlier options, and because if there is a clear or nuanced departure from the customer requirements, that departure must be identified.

The legal purpose is obvious and has been discussed in Chapter 4. James Kolka, an attorney in international standards and requirements, writes that a paper trail of responsible design management provides a reasonable defense in the event of litigation following the design.

A.2   Forensic Analysis of the Design Process

There is an old saying: "Fools rush in where angels fear to tread." This admonition is often on the mind of forensic systems analysts who find themselves analyzing an activity in which they may not be experts. Certainly, auditing a complex system of product or service for compliance is an intimidating prospect, and the design process is as complex as most. This is why an analyst is well advised to understand the process shown earlier in Figure A.1. You do not need to understand all the details of each block, but you must be able to ask the right questions, and understanding the flow diagram of the figure helps you to follow a comprehensive and reasonable pursuit of the design process in litigation.

In keeping with the policy of using ISO 9001:2015 as a standard of operations, Clause 8.3 of that document, *Design and Development of Products and Services*, will serve as our guide. The following outline can help you in formulating your questions.

- Customer requirements must be determined and translated into specifications. This includes the determination and timely availability of materials, skills, information (drawings, software, policies, procedures, technical manuals, safety issues, statutes, regulations, and environmental constraints), and other necessary resources.
- A preliminary set of drawings represents the initial design and the process is iterated in a development loop until an optimal design is achieved in terms of the final requirements. This loop includes developing prototypes or configurations, with verification and validation procedures leading to stages of assessment, amendment, and improvement.
- Reviews are conducted for each development stage, to include the participation of all interested persons, for the purpose of considering issues of assessment. The reviews are open to discussion of improvements in the design of the product, service, processes, or even of the specifications.
- Acceptance testing is conducted of the final design configuration. This testing is itself designed as a verification and validation of the designed product or service.
- Records are maintained of all review results and of all changes. These records are treated as quality records and include the updated final drawings, specifications, and approvals. The customer is the primary approval agent of record.

The integrated processes of Figure A.1 enhance design sustainability with three key cycles:

1) requirements ↔ specifications
2) specifications ↔ prototypes
3) prototypes ↔ acceptable design

In addition, the methodology for each cycle has three parts: verification, validation, and participation by external and internal customers. The external customers include buyers and other interested persons. The internal customers represent the performer's employees and management with cross-functional expertise in process, product, marketing, purchasing, and customer service.

References

Arora, J. S. (1989). *Introduction to Optimum Design*. New York: McGraw-Hill, pp. 4–11.

ANSI/ISO/ASQ (2015). *ASQ/ANSI/ISO 9001-2015: Quality Management Systems—Requirements*. Milwaukee, WI: American Society for Quality and the American National Standards Institute, pp. 11–12.

Gibson, J. E. (1990). *How To Do Systems Analysis*. Charlottesville, VA: University of Virginia School of Engineering and Applied Science workbook.

King, R. (2003). *Michelangelo and the Pope's Ceiling*. New York: Walker Publishing, pp. 45–70.

Six Sigma (1998). Charlottesville, VA: GE Fanuc Automation, p. 24.

Stimson, W. A. (2001). *Internal Quality Auditing: Meeting the Challenge of ISO 9000*. Chico, CA: Paton Press, p. 180.

Appendix B

Introduction to Product Reliability

In the Preface, I presented the ISO 9000 definition of product quality as *the degree to which a set of inherent characteristics of a product or service fulfils customer requirements.* While speaking of the quality of a product or service, this definition says nothing about how long the quality will last. What is its durability? What good is the quality of the thing if it is gone by the time you get home?

Some products are meant to be disposable and their durability is not critical. Others might be expected to last a long time, such as appliances. The durability of some products is marked on the package, light bulbs being an example. I use the word *durability* because it is generally understood to refer to how long a product can be useful and the term is often used as a product characteristic. However, durability is not a technical term and has no meaning to engineers. It means whatever the manufacturer wants it to mean.

Forensic Systems Engineering: Evaluating Operations by Discovery,
First Edition. William A. Stimson.
© 2018 John Wiley & Sons, Inc. Published 2018 by John Wiley & Sons, Inc.

When referring to how long a product can meet intended requirements, engineers use the term *reliability* and define it rigorously (Grant & Leavenworth, 1988):

> Reliability is the probability that an item will perform a required function without failure under stated conditions for a specified period of time.

The beauty of this definition is that once the operating conditions are stated and the life time estimated, the definition of reliability becomes a specification.

Reliability, then, is quality that endures. Today, we often say that *reliability is quality over time* (Lochner & Matar, 1990). However, you do not get it for free. Reliability is a characteristic of a product and must be designed into the product or its existence is problematic. In addition to design costs, reliability must be tested in the lab during the development phase and further verified with warranty data from the field after sale, when the product is in use. All these actions add to the cost of production.

B.1 Reliability Characteristics

B.1.1 Reliability Metrics

Product reliability is verified by measurement. The metric for products that cannot be repaired, such as light bulbs, is *mean time to failure* (MTTF). The metric for products that can be repaired, such as digital computers, is *mean time between failures* (MTBF). Some products are designed for a single use, such as airbags, missiles, and body armor. The metric for single-use products is *probability of failure on demand* (PFD). These metrics are in reference to a period of time. The inverse of a period of time is its rate. The failure rate of a product is an additional metric of frequent use in reliability engineering and you will see it used from time to time, but it is simply another form of the same information.

Clearly, reliability is a statistical parameter. *Mean time* refers to the average lifetime of a population of products. You cannot talk about the reliability of a single product except to compare it to the reliability of the population it is drawn from. As with any average, some products will fail sooner and some later than the specified time. So if you are a buyer of products in quantity and expect a certain reliability of that quantity, you must track the average performance life of the products that you have purchased. If this average life is approximately that specified by the manufacturer, then you are getting what you paid for.

There are a number of ways to formulate reliability, but the most common one is an exponential model:

$$R(t) = e^{-\lambda t} \tag{B1}$$

where λ is the failure rate, often measured in failures per hour. When the failure rate is constant (the population of the units is in a state of random failures), then $\lambda = 1/\mathrm{MTBF}$.

There are many who claim that in the physical world a constant failure rate does not exist. Perhaps not, but nature comes close; close enough for engineering purposes. There is a very large body of evidence of product failure over the years that shows the utility of reliability predictions. And if there were no approximately constant failure rate in the physical world, then all marketing would be a false claim. No one buys a product without some assurance that the product will last awhile.

Moreover, mathematical analysis of reliability is relatively easy when failures are modeled as random and hence, product life analysis is usually focused on that period of its life cycle when the failure rate is fairly constant relative to early and late product life. The methodology of reliability does recognize two nonrandom states: infancy and old age, but these states prove to be transient. Statistically, there is an effective state of random wear in the life of a product; it is a stable phase that can be defined by a constant failure rate. In material life, as in human life, infant mortality and old age (or wear-out) are not stable phases. They require special attention.

B.1.2 Visual Life Cycle

All products deteriorate with time. Various models have been used to describe this deterioration, among them the "bathtub" model shown in Figure B.1 (NIST, 2012). The bathtub shape describes three distinct phases of a product life cycle: an initial decreasing failure rate, called infant mortality; a constant failure rate, called the useful life of the product; and an increasing failure rate, called wear-out. An argument is made that with the advent of relatively long-lasting semiconductors and the rapid pace of technological obsolescence, the bathtub

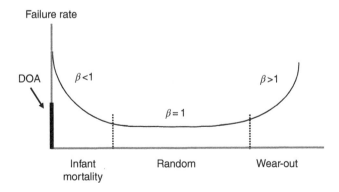

Figure B.1 Product life cycle phases (Bathtub model).

curve is no longer relevant. The riposte to this criticism is in two parts. (i) Modern large systems are composed of many elements, including humans, each with its own reliability and fragility. (ii) The bathtub curve is an integration of three demonstrably active responses in the life of a product: an initial failure rate; a steady state failure rate indicating the designed life of the product and that must exist if the product is to have any marketability at all; and a final increasing failure rate as the product begins to wear out. It seems to me that the bathtub curve will be useful to forensics in the general case for some time to come, irrespective of technology.

For the moment, the reader should ignore the β notations in the figure and simply focus on the shape of the curve. The initial phase begins when a product first comes to market and may have a disappointingly high rate—indeed, the very first event may be that it fails to work at all. This is called "dead on arrival." The causes of infant mortality are usually from defects or errors in manufacturing, shipping, or installation, but rarely from poor design if reliability testing had been included in the design phase.

Infant mortality may also be caused when a defective unit on the production line is reworked, but then it is not retested. Many failures in the field are caused by reworked products that had failed in-process test or inspection. Infant mortality may also be caused by an ineffective test program or because the product was released to the market before being fully tested. Premature release is usual in the software world and is called Beta-testing. The producer does some product debugging but argues that it is not possible to predict all possible uses and abuses by the customer, a certain amount of debugging remains. On occasion, a customer may demand early product release for operational reasons and will accept the subsequent infant mortality. The infant mortality rate can be modeled by a Weibull distribution with a relatively large rate of decay.

Although there is some overlap between phases, at some point "infancy" fades away and the failure rate steadies out, forming the flat portion of the graph. This period is called the normal life of the product and failures occur randomly during this time. The notion of a constant failure rate is important because it permits mathematical derivation of life cycle characteristics. The mean life of a product is estimated from experiments or from field data and provide the failure rate, λ, shown in Equation E.1, that is applicable to this steady state, if you will. This phase of the product life is modeled by an exponential distribution, found to provide the best model to approximate arrival intervals, being the time between failures (MTBF) or from start-to-first failure (MTTF).

Eventually, wear and tear become significant factors and the product enters into a phase in which wear-out becomes increasingly dominant. Causes of wear-out can be frictional wear, fatigue, corrosion, chemical changes, shrinking, cracking—a multitude of events that we attribute quite simply to old age. The failure rate increases rapidly. This phase can be modeled by the normal distribution.

B.2 Weibull Analysis

B.2.1 Distributions

In 1939, a Swedish engineer named Waloddi Weibull introduced a methodology that eventually became one of the most widely used tools in reliability engineering for estimating product lifetime. Weibull analysis matches failure data to distributions that represent the failure rates of products under study, whatever their phase in the life cycle. Given a set of failure data, the underlying distribution can be determined.

Weibull distributions are generated by a probability distribution function that is rather cumbersome for the purposes of this book; however, the Weibull cumulative distribution function (CDF) is much simpler and quickly understood:

$$F\left(t,\beta,\eta\right)=1-e^{-(t/\eta)^{\beta}} \tag{B2}$$

Hence, Weibull creates various distributions by varying two parameters: the *Shape*, β, which gives the form of the distribution and the *Scale*, η, which gives its spread or range. The shape is often called the *Weibull slope* because it can be graphed as such. When discussing Weibull studies, the reader may find that shape and slope are interchangeable names for the same thing.

Scale indicates the characteristic life of the thing measured and is the point in time where 63.2% of the products have failed (McLinn, 2009). This is called the *characteristic life* of the product and represents, approximately, the mean life, be it MTTF or MTBF. (There is a third parameter, "offset," that is required if the failure data do not all have the same starting point, but this complication is unnecessary to this book. Our purpose here is to enable the reader to recognize the evidence of reliability studies, if such exists.)

Each of the distributions shown in Figure B.2 tells a different story. The leftmost curve, $\beta = 0.5$, is an example of infant mortality ($\beta < 1$) and describes an initial failure rate that diminishes with time. The curve, $\beta = 1$, represents the normal product life that is described by a constant failure rate. However, it too begins to diminish as the number of units diminishes (see Section B.5). The MTTF and MTBF are calculated relative to the number of devices. As the number diminishes, so also will the failure rate diminish.

The curve, $\beta = 2$, is an example of the early wear-out phase ($\beta > 1$) and describes a linearly increasing failure rate that grows with time, but then begins to diminish as the number of units diminishes. The center most curve, $\beta = 4$, is an exponentially increasing failure rate. It is represented by a normal distribution, describing a rapidly increasing wear-out rate that then decreases as the population decreases. Each of these, and many more, can be created from varying values of β and η.

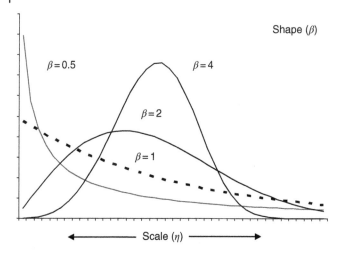

Figure B.2 Examples of Weibull distributions.

Weibull analysis is particularly useful when the product is in its phase of constant failure rate—the flat part of the bathtub curve. During this phase the failure distribution is exponential and mean life can be easily determined. (It should be noted that the mean life only has meaning when the product is in its useful phase). The exponential distribution is not symmetric and its mean, the expected product life, indicates that 63.2% of products will have failed. This is an awkward number and not particularly popular. I shall comment on this later in this chapter.

The Weibull expression for product reliability is simply the CDF less unity:

$$R(t) = e^{-(t/\eta)^{\beta}} \tag{B3}$$

When $\beta = 1$ the failure rate is constant. Then comparing Equation B3 to Equation B1, we can conclude

$$\lambda = \frac{1}{\eta} \tag{B4}$$

which agrees with our earlier finding: $\lambda = 1/\text{MTBF}$. Hence, when the failure rate of a product is constant, the MTBF and the characteristic life are equal. These analyses are appropriate to services also.

B.2.2 Shape and Scale

B.2.2.1 Shape
Examining Equation B2, it is clear that both the shape and the scale will affect the resulting distribution. Figure B.2 shows how the shape parameter, β, can

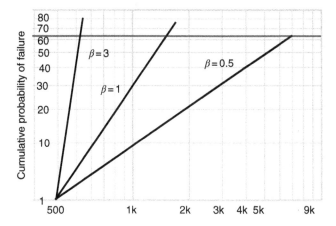

Figure B.3 A Weibull probability plot.

completely change the character of the distribution. However, its effect in assessing failure data is also powerful and indicates why the shape parameter is often called the Weibull slope. Figure B.3 is a plot of several lines representing different values of β. These straight lines are typical of those generated by the method of least squares on data of failures plotted on logarithmic paper.

The importance of such a graph is that, given a set of failure data either from the lab or from the field, the slope formed by the data will indicate both the β value and the η value. The first reveals the underlying distribution; the second provides the characteristic life.

If the data are from the lab, they verify the design; if the data are from the field, they validate the product. Later in this appendix an example of verification is shown, with descriptions on how to gather the information that the graph offers.

Refer back to Figure B.1 to see how the life cycle curve correlates to the Weibull parameters. If the data show β to be less than unity, we know the product is in the early phase of its life—the infant mortality phase. If β is greater than unity, the product is in the wear-out phase. $\beta = 1$ marks the flat part of the curve. This means that the failure rate is constant or nearly so and failures are random. A constant failure rate greatly simplifies the mathematics of reliability, for when $\beta = 1$, $\eta = \text{MTTF}$, so:

1) The MTTF is the reciprocal of the failure rate, say λ ($\text{MTTF} = 1/\lambda$). However, when β is greater than unity, the MTTF becomes less predictable and as β becomes much greater than unity, the MTTF no longer applies. All bets are off.
2) When $\beta = 2$, there is a linearly increasing failure rate, and as β increases from there, the failure rate increases exponentially.

As a forensic systems analyst, you may not be interested in the mathematics of reliability, but it is to your advantage to recognize the "artistic" features of Weibull—its charts and graphs and what they say about where a product is in its life cycle. The reliability of a product may be a significant issue in litigation.

B.2.2.2 Scale

At this point, the reader might wonder why the scale, η, is needed when we have the MTTF? In reliability studies, Weibull analysis is used to find the distribution of failure data. The scale, η, is an essential part of that process and marks the 63.2% accumulated failure estimation. To Weibull, η is the best estimate of product life because it is the mean value of an exponential distribution that indicates the useful phase of product life. But to most people, 63.2% is not a very handy number to remember or to understand. The 50% point is much easier for nontechnical persons to remember and is more intuitive. Fortunately, Weibull has simplified this problem for us with the B-percentile concept.

B.2.3 The B-Percentile

Reference to the B-percentile may be found in discovery and is easy to grasp if you know the technical vocabulary of reliability, but it is very confusing if you don't. The x-axis of a probability plot usually references the metric of interest, be it distance, time, or some other, and is paced off in units of standard deviation. For purposes of lifetime estimates, Weibull changed this spacing, preferring to use percentiles of cumulative product failure (Abernathy, 2006). Thus, he defined points such as "B0.1" and "B50." B0.1 would be that value of metric that corresponds to the 0.1% cumulative failure intercept. B50 would be the value of the metric that corresponds to 50% cumulative failure intercept, or roughly the MTTF.

In Figure B.4, these points are mapped onto a normal distribution in which the x-axis is expressed in both the usual standard deviations from the MTTF and the percentile value. This allows the integration of traditional quality ideas to those of reliability. For example, -3σ describes the value of the left tail at 0.00135. In percentage, this would be 0.135% and values less than this fall into the category of rare event.

Notice that the left tail of the distribution is very close to Weibull's B0.1, as shown here, which defines the 0.1% cumulative failure. In fact, in some industries they refer to the B0.1 point as the -3σ point. They do this in an attempt to align the terminologies of Weibull with process control. Thus, B0.1 defines a rare event.

Similarly, the peak of the normal curve coincides with B50. This statistic is of interest because it represents the mean value of product life. So when you see B0.1, then think "-3σ." When you see B50, think MTTF. Finally, B10 is also a point of interest to those who use the B-axis. It corresponds to -1.2σ in standard deviations and represents the time when 10% of the product volume has failed.

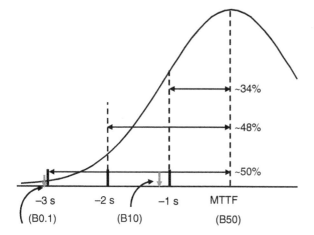

Figure B.4 Examples of the B-percentile of product life.

B.3   Design for Reliability

Reliability, as with every key characteristic of a product or service, must be designed into the unit if it is to exist. Therefore, it is incumbent on the forensic systems analyst to be familiar with the design process in general terms, so as to recognize from the evidence whether such a process has been used in the subject in question. Therefore, the reader might consider perusing Appendix A, which provides an easy-to- understand review of the design process used in engineering. All of the phases of design from initial planning to acceptance testing apply to the design of reliability. Moreover, there should be a paper trail through each of these phases.

The terminal event of reliability is failure. Product liability is keyed to the period of acceptable performance and the purpose of reliability design is to extend this period. There will be many occasions for doing so, one at the initial design of a product and then iteratively during the product life cycle, depending on how it performs in the field.

An essential technique of reliability design is called *Failure Mode and Effects Analysis* (FMEA), an anticipatory strategy used to estimate product reliability. FMEA takes a step-by-step approach to identify all possible modes of failure in a design, a manufacturing or assembly process, or in operations. In a brainstorming environment, a working group of diverse experts asks "how can a product fail?" then estimates the probability and severity of that failure mode. They then determine design features to reduce the probability or the effect of occurrence, focusing especially on the risks with severe adversity.

An Ishikawa diagram (example shown in Figure B.5) is a useful tool in a brainstorming session. It is named after its inventor, Kaoru Ishikawa, but it is equally well known as a fishbone or cause-and-effect diagram.

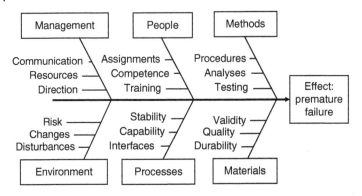

Figure B.5 A cause-and-effect diagram.

Premature failure is an effect due to common and special causes. Common causes are random and always exist because no product is perfect. Special causes are not random but derive from conditions external and foreign to the product's design. Design engineers get together and try to hypothesize how a product can fail.

Causes of a problem can be classified in several general categories, usually selected in accordance with the nature of the problem under study. Focusing on the system perspective, we have selected major categories listed as management, people, methods, processes, materials, and environment. Within these categories, brainstorming develops specific causes that may vary greatly, depending on the product or process that failed. In Figure B.5, several typical causes are shown within each category, just to indicate how an investigation into product or process failure might be initiated. Cause-and-effect diagrams are equally useful in the design phase of a product or service to anticipate where failure might occur. They are also used after a given failure in the field to trace its cause. Such diagrams are also part of the paper trail of evidence of design of reliability.

Many of the factors that affect the reliability of a product derive from its quality. This is no surprise because reliability is simply quality over time. You begin with the initial quality, defined by fitness for use and design for robustness—the ability to remain fit when subject to expected disturbances. The character of robustness suggests that the driving forces of failure are major categories and the required factors as minor categories to that end. These ideas begin the construction of the diagram. At this point, open discussion is essential. Outside-the-box thinking is encouraged. The durability of the material quickly comes to mind, but no possible factor should be overlooked, that is, the symbiosis of interfaces or of the environment.

Durability requires analysis of the aging properties of the material and its ability to withstand stress. In this context, stress refers to all those factors that act adversely on the product. Ruggedness describes how well the material maintains its condition under expected use.

Products usually pass through several processes whose stability and capability are critical to the quality of the product. The effectiveness and efficiency of the interfaces between processes also play a critical role in the composition of the product, whose chemical or other inherent properties depend upon continuity of process.

The environment, from a system's perspective, is all that which may affect the system output but which are not part of the system. The major exogenous forces are the risk of adverse events, disturbances in the processes or changes in any critical issue relevant to product quality: material, time, craftsmanship, heat, refrigeration, or any necessary environmental characteristic. Risk is a probability of an adverse external event and hence refers to the possibility of a change or disturbance; therefore, its consideration is an essential component of the planning needed for the design for reliability. A product must be designed to perform as required over a specified range of all foreseeable factors and to hold up over a specified period. Operational limits must be understood and agreed to by both buyer and seller. Misuse by the user may void both warranty and reliability.

The brainstormers proceed through the major categories of the diagram, trying to imagine all relevant categories and the things that can go wrong. Clearly, the group should have diversity of experience, representing both the top-down and bottom-up views of system performance. Design records may indicate the width of expertise employed in the risk analysis conducted in corporate FEMA activity.

B.4   Measuring Reliability

B.4.1   On Reliability Metrics

There are many ways to measure the reliability of a product, but the most common one is simply to measure how long it lasts. As the lifetime will vary from product to product, the mean value of product life is measured with metrics as described earlier. For example, when you buy a light bulb, the MTTF is printed on the package. The MTTF is the easiest measure to calculate; a simple arithmetic average usually provides a good estimate, with this caveat: the mean life time is not hours per failure, but the number of devices multiplied by the hours per failure. For example, the MTTF of one device that lasted for 100 hours is the same as the MTTF of 10 devices that lasted for 10 hours.

During the useful phase of a product's life, the MTBF is calculated similarly to that of the MTTF. Because of this, it should be remembered that they are not the same thing and are used for different purposes. For failed products or systems that can be restored to service, the MTBF may be recorded. As an example, the US Navy often requires that an internal clock be provided in an equipment for this purpose. The internal clock tracks the operational time, which should be recorded daily.

The rate of failure of a product, noted as λ in this chapter, must be specified over a period of time. This rate is quoted in various ways such as "failures per 1000 hours." If the failure rate is constant, the product MTBF can be easily estimated from it by taking the reciprocal of the rate:

$$MTBF = \frac{1}{\lambda} \qquad (B5)$$

The difference in kind between MTTF and MTBF can be appreciated by looking at actual data that may appear in discovery. Large systems contain many components, each with its own reliability. Very often, but not always, when the components fail they are discarded. Thus, their reliability is measured by MTTF. The system itself, perhaps very expensive, is repaired and not discarded. Its reliability is measured in MTBF.

Consider a high-pressure turbine blade with an MTTF of 9000 hours. About half of them will still be performing as designed after 9000 hours of operation. Yet, the MTBF of a jet engine is considerably less than that—700 hours (Zaretsky & Hendricks, 2003). This is because the engine has many components and if they independently fail, the engine failure rate is the product of individual component failure rates. This means that maintenance schedules will be keyed to the jet engine life cycle first of all. Then during a given engine maintenance period, appropriate replacements are made of the various components. For example, in the case of an engine with an MTBF of 700 hours, its maintenance schedule may call for an overhaul of, say, every 500 hours. At that time, engine components will be examined for fracture, corrosion, fatigue, and other signs of wear and tear, and refurbished or replaced as needed.

A similar relationship exists with radar systems and computer systems. A given radar transmitter may have an MTTF of 46,000 hours, but the radar system of which it is a part may have an MTBF of only 1100 hours (Young, 2003). The hard drive in a computer system may have an MTTF of one million hours, but the computer system itself may have an MTBF of 1700 hours (Schroeder & Gibson, 2007).

These descriptions indicate the relationship of the reliability of systems and of the components of those systems. We are interested in this relationship because this is a book about systems. However, the reader should not assume that this relationship defines the metrics. MTBF and MTTF apply to repairable and non-repairable units whether they are related or independent; operated individually or conjointly as a system component.

B.4.2 Graphing Failure Data

The quality of a product design depends upon the performance of units in the field and this is true of the design reliability also. Given that reliability was

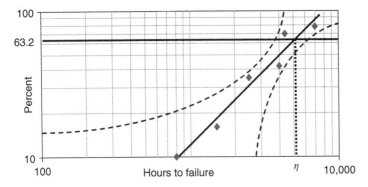

Figure B.6 A Weibull graph of time to failure.

indeed designed into a product, its true reliability can be verified by failure data occurring in use, usually called field data. Figure B.6 shows a typical Weibull graph of failure data. If you are involved in a case in which reliability is a factor, this kind of graph may be important evidence one way or the other. A Weibull graph is viewed from two perspectives—macro and micro. Let us examine the macro view, or big picture, first.

The macro view in Figure B.6 shows a straight diagonal line, two curved lines, a horizontal line at the 63.2% mark of the vertical scale, and horizontal and vertical grids. In addition, there are several points of product failure data. The slope of the diagonal line, called the *regression* line, tells the story—it represents a best estimate of the probable failure history of this product. It is drawn from the six points of actual data by the method of least squares fit, which means it is the line such that the distance from the line to any one of the data points is as small as possible. It is necessary that the points actually indicate a straight line on a logarithmic scale and if so, the underlying distribution of failure can be determined. If the data do not define a straight line, there may be an offset in the time to failure or a mix of failure modes, and the underlying distribution cannot be established, at least not from the available failure data.

The slope of the regression line reveals the shape of the underlying distribution of the random failure points. Knowing this distribution, we can then calculate confidence intervals about the regression line. The slope of the line in this graph, β, is given as 1.6, which indicates that the product is in its early wear-out phase. To see this, refer to Figure B.1, which shows that the wear-out period begins at $\beta > 1$. Conversely, $\beta = 1$ indicates that the product is in its normal life, subject to random failure. Hence, the slope of the regression line on log paper indicates where the product is in its life cycle.

The confidence region of the graph is defined by the curved lines and represents the statistical confidence we have that the data do indeed reflect the probable life of the product. A 95% confidence level was selected for this test.

The lines are oriented parallel to the regression line and form a confidence interval that varies in their distance to it, depending on the degree of confidence. Near the cluster of data the interval narrows, indicating more confidence that the interval contains the mean. This is because the cluster provides more information about the reliability in that span of time. Conversely, the farther away from the cluster, there is less information about the reliability of the product and the confidence interval widens, as we are less sure of when failure will occur.

The horizontal grid line at the 63.2% accumulated failure level gives the reference to the scale of the underlying distribution at the point where it is intercepted by the diagonal line. The scale, η, is 5200 hours, which you find by simply dropping a plumb line down from the intersection of the β-slope with the reference line.

How well does the regression line estimate the product time to failure based on the samples of failure data? This "goodness of fit" can be measured by the correlation of the data and turns out to be 0.9489, when run through EXCEL. This means the regression line accounts for 94.89% of the variation in the data. This is a very good accounting, but it should be understood that the correlation measures how well the line fits the data, and not how well the data tells the true history of product failure. Whether a sample size of six is sufficient to indicate the population life depends upon how much we already know about the history of the product.

Now consider the graph from a micro view. The regression line and confidence curves represent an infinite set of points that connects the cumulative failure, in percent, to a unit of time. For example, suppose you want to know the B50 estimate of this product—the time in which 50% of all the products put into use at an initial time will fail. You go to the number "50" on the vertical axis and follow it over till it intersects with the regression line. Then you drop a plumb line down to the horizontal axis and read the estimated time. It is about 4000 hours.

Yet, this line was drawn from sparse information—we had only six failure data. How confident can we be in this 4000 hour estimate? To answer this question, you choose a desired level of confidence. In this case, the CL is 95%. So you go again to the number "50" on the vertical axis and follow it over till it intersects with the first confidence line. Drop down to the horizontal axis and record about 3300 hours. This is the low estimate of the 95% confidence interval.

Then continue again along the "50" line to the second confidence line and drop down. The high estimate is about 5000 hours. Thus, you can say with 95% confidence that 50% of this product line will fail as early as 3300 hours and as late as 5000 hours. That is quite a range of uncertainty, but we have only six points of failure data. If there were more failure data available, the confidence interval would be tighter.

B.5 Testing for Reliability

Testing and measurement are closely related—you cannot perform a test without making measurements—but for forensic purposes I distinguish the two. Measurement is the comparison of two values, one of them a reference. A test is a procedure for determining how well something works or determining the truth about something.

A test will require a plan, an objective, formal procedures, and consideration of various stimuli and responses. Testing is the act of verifying or validating an entity against defined criteria for some purpose. You verify and validate by measurement and this unites the two activities. Verification is the act of ensuring that the unit under test conforms to specifications. Validation is the act of ensuring that the unit under test conforms to its intended use—that it is what the end user really wants.

Product life can be measured either with field data or laboratory data gained through experiment. Field data are the actual failure data coming in from products in use, such as casualty reports, field engineering reports, customer-gathered data, and warranty data. Warranty data are an especially good source because they can be used to measure the true reliability of a product.

The problem with field data is obvious—it is after the fact information. The product is already designed, fabricated, and in use. Field data are necessary to establish actual product reliability, but the customer deserves some indication of its reliability before the purchase is made. You need an *à priori* estimate.

Laboratory experiments provide the data for an *à priori* estimate. If the expected product lifetime is relatively short or the product is inexpensive and can be tested in sufficient numbers, you just operate them in the lab in real time until they fail. This is called *life testing*. The ratio of the number of survivors to the total tested over a given period of time provides an estimate of product reliability.

Much of reliability testing is run to fail. Useful information about the design can be gained by measuring how long the product will last and why it failed. A life test should aim for at least three failures. One failure will not indicate variability and two failures will not indicate nonlinearity. We discussed Weibull analysis in Section B.2. As with many kinds of measurements, it is used extensively in reliability testing.

Life testing, as its name implies, is designed to test the operational life of a product. The concept of a life test is straightforward. A number of units are run until they fail and then the failed units are counted with respect to the total test population.

Suppose that a manufacturer wants to estimate the failure rate of a microchip over a 1000 hour period. A statistically significant number of chips are selected for operation, say 100, then the group is run for 1000 hours. At the end of the period, suppose that three have failed. If the products are in the constant failure rate of their life, then a valid estimate can be made for the MTTF of the microchip. Bear in mind that although the test is 1,000 hours long, 100 chips ran for this 1,000-hour period, so that effectively we have 100,000 hours of chip operation. The MTTF is based on this total operation.

As three chips failed of the 100 tested, the unit failure rate is 0.03 or 3%. However, the MTTF is based on total operational time, so with respect to this time, the failure rate is three chips per 100,000 hours. Thus, the MTTF is the reciprocal of the total failure rate, or 33,333 hours, which is roughly 4 years.

If the expected life precludes real-time testing, then you need accelerated testing. This is achieved by increasing the stress level on a product. For example, you might increase the operational temperature and humidity beyond the expected use requirement. There are various models to relate product life to stress level and choosing one of them depends on the expected dominant change factor: heat, humidity, metal fatigue, operating voltage, and so on. Other than to give an example, I shall not delve into accelerated testing, which requires a very deep understanding of the properties of a product, as well as a broad and experienced background in reliability testing methodologies.

The basic assumption of an accelerated life test is that a product, operating under higher levels of stress, will experience the same failure mode as seen when used at a normal stress level. The only difference being that the failures will "happen faster." Depending on the product to be tested, stress can be temperature, electrical voltage or current, humidity, pressure, shock and vibration, load, cycling or use rate, or other influencing factor.

Figure B.7 is an example of an accelerated test of heat sinks. Heat sinks are used in high-performance computers to carry away heat from the central processing unit, but over time they form internal corrosion that degrades their ability to transfer heat. Assume that the upper operating temperature is specified at 40°C. The heat sink must control ambient temperature such that 40°C is

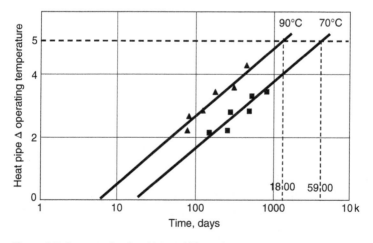

Figure B.7 An example of accelerated life testing.

not exceeded. If the ability of the sinks to carry away heat is such that the internal operating temperature increases by more than 5°C, they are considered unusable. The required MTTF of the sinks is specified to be 10 years.

However, 10 years is an unrealistic period in which to conduct a continuous test of reliability. Assume that engineers determined the dominant stress is ambient temperature. A dozen sinks of specified composition were tested to failure over a 300-day period, in two groups of six each and at two levels of excessive temperatures: 90 and 70°C. The critical baseline was a five-degree increase in operating temperature.

The failure data at different temperatures maintained the same slope, with only the scale changing. This result means that the failure modes did not change characteristics under stress but simply failed at different times. Thus, the sinks will operate effectively at a temperature of 70°C for at least 5900 days, or 16 years. This more than meets the specification because as the specified operating temperature is 40°C, considerably lower than 70°C, we can expect a greater MTTF than 16 years.

References

Abernathy, R. B. (2006). *The New Weibull Handbook*, 5th ed. North Palm Beach, FL: Self-published, p. 5. http://www.barringer1.com/pdf/Chpt1-5th-edition.pdf, Accessed August 5, 2016.

Grant, E. L. and Leavenworth, R. S. (1988). *Statistical Quality Control*. New York: McGraw-Hill, p. 582. A similar definition is provided by the American Society for Quality at http://asq.org/learn-about-quality/reliability/overview/overview.html. Accessed November 29, 2015.

Lochner, R. H. and Matar, J. E. (1990). *Designing for Quality*. New York: Quality Resources and ASQ Quality Press, p. 4.

McLinn, J. (2009). *The Weibull Function and How to Understand It*. Weyers Cave, VA: Blue Ridge Community College and the American Society for Quality. Industrial course (February 13, 2009).

NIST (2012). NIST/SEMATECH e-Handbook of Statistical Methods. National Institute of Standards and Technology. http://www.itl.nist.gov/div898/handbook/apr/section1/apr124.htm. Accessed July 31, 2016.

Schroeder, B. and Gibson, G. A. (2007). *Disk Failures in the Real World: What Does An MTTF of 1,000,000 Hours Mean To You?*. Pittsburgh, PA: Computer Science Department, Carnegie Mellon University.

Young, R. B. (2003). *Reliability Transform Method*. Blacksburg, VA: Virginia Polytechnic Institute, p. 207.

Zaretsky, E. V. and Hendricks, R. C. (2003). Effect of Individual Component Life Distribution on Engine Life Prediction. *NASA/TM—2003-212532, 2003*, Glenn Research Center, Cleveland, OH & United Airlines, San Francisco, CA. pp. 10–11.

Appendix C

Brief Review of Probability and Statistics

Statistics has been defined as making sense out of data. The two major branches of statistics are *probability* and *statistical inference*. Probability is a methodology used to describe the random variation of data; in our case, systems data. The range of values in a set of data may vary and we may not know its average value exactly if the population we are examining is very large. For example, we cannot know the average number of salmon in the Columbia River in any given year. Nor can we know exactly the average value of a product made by a company as long as it continues making them, because all the while that production continues, the population is changing values.

The average value and standard deviation of a random variable are called parameters. Statisticians use a convention to distinguish the true values of the parameters, which we often do not know, and the measures taken of them. True values are labeled by Greek symbols. The true mean value of a set is often labeled by μ; the true standard deviation of the set is always labeled by σ. Measured

Forensic Systems Engineering: Evaluating Operations by Discovery,
First Edition. William A. Stimson.
© 2018 John Wiley & Sons, Inc. Published 2018 by John Wiley & Sons, Inc.

values are labeled with letters of the alphabet. If a parameter is continuous, then parametric estimates are always labeled $\bar{x}$ for the mean and s for the standard deviation. If the parameter is discrete, say the fraction rejected of a population, its parameter estimates may be p for the mean and, in this book, s for the standard deviation.

We take samples and determine their average value and deviation. If we do this correctly and often enough, the average of the sample averages approaches the true value and we can make very good estimates of the random properties of a stochastic process. These estimates are called statistics.

In measuring random data, you need a metric that defines the essence of what you are measuring. A metric is the unit of measurement you choose to measure something with. For example, you can choose to measure speed in feet per second, miles per hour or even furlongs per fortnight. You can measure length in centimeters, meters, or miles.

The key characteristics of any product or service are stochastic—they vary randomly in time. Therefore, forensic investigation of a data set descriptive of such operations will require a familiarity with the fields of probability and statistics.

Statistics is a rather large field of mathematics, but for purposes of this book we only need to learn a few ideas. In particular, we should know the measures of location in a random activity, the measures of dispersion of that activity, and we should know what the word "distribution" means in statistics. In forensics, it is important also to understand how to measure extreme values of an activity.

C.1 Measures of Location

When numerical data are taken, as in a measurement, the data tend to cluster about certain numerical values. This is called the *central tendency of measurements* (McClave & Dietrich, 1988). The average value of a set of data is a value that is descriptive of the set. The three most common types of averages used in statistics are the mean, the median, and the mode. The kind of average to be used in a given situation depends on the objective of the measurement, but usually the choice is clear—you choose to measure the average that best represents the set of values in line with your objectives. On occasion you may want to know two or all three of the averages, because each has a story to tell.

C.1.1 Average: The Mean Value

The mean value of a set of data is the sum of all the numerical data divided by the number of data points. For example, assume that there are n values in a set, each value represented by x_i. Then the mean value, $\bar{x}$, is found by the following equation:

$$\bar{x} = \frac{1}{n}\sum_{i=1}^{n}x_i.$$

(C1)

The mean value is the most commonly used measure of central tendency. In a discussion of probability distributions, the mean value is used for symmetrical or near-symmetrical distributions, or for distributions that lack a clearly dominant peak (Russell, 1997a). The advantage of the mean as a measure of the average value of a set is that it includes all the values in the set. Conversely, the disadvantage of the mean is that it is influenced by outliers even when they may totally misrepresent the set. For example, the county I live in is rural and generally populated by middle-income homes but with a scatter of small, poor farms and several vast estates. The difference in value between the small farms and the middle-income homes may range in the thousands of dollars, but the difference between the middle income and vast estates ranges in the millions. The mean value of properties in this case could very well indicate that the county is more wealthy than it really is, therefore affecting Federal and state ideas of taxation.

C.1.2 Average: The Median

The median value of a set of data is the value at the center of the data when they are ordered by magnitude, either ascending or descending. If there is an even number of values, then the median is found by taking the mean value of the two center values. For example, if there are 19 values in a set, then the median will be the 10th value. If there are 20 values in a set, then the median value will be (10th + 11th)/2.

The advantage of the median as a measure of the average value of a set is that it is not influenced by outliers, on the contrary, it can be used to reduce the effect of extreme values. Suppose that you are looking for a home in a good neighborhood, but you have an upper limit of $300K. Two neighborhoods have houses for sale with prices as shown:

A) 243K, 248K, 256K, 262K, 269K, 278K, 304K.
B) 208K, 215K, 235K, 262K, 265K, 592K, 836K.

Neighborhood A has a mean value of $266K, clearly in your range. However, the mean value of Neighborhood B is $373K, out of your range. Yet, both neighborhoods have the same median price of $262K. If the real-estate agent reports the median values to you, then your opportunities are doubled.

C.1.3 Average: The Mode

The mode of a set of data is the value that occurs most frequently. If all values are different, then there is no mode. If two values have the most representation but the same frequency of occurrence, then the data set is *bimodal*. The mode may be used for severely skewed distributions, for describing an irregular situation in which several peaks are found, or for excluding extreme values (Russel, 1997b). An irregular situation refers to a bimodal distribution resulting from a mix of two different unimodal distributions caused by a heterogeneous population. In manufacturing for example, this would indicate a single system with two distinct but perhaps unknown behaviors. This latter condition could be of interest in forensic investigation because stable systems are well behaved.

C.2 Measures of Dispersion

In random variation, data are spread about an average mean value, most of the dispersion in the area of central tendency. This dispersion is called variation and less often, scatter. If the variation is in reference to the key characteristic of a process output, intuitively we would say that the less, the better. Measurement of the dispersion yields vital indications of process stability and its use is essential in descriptive statistics and hypothesis testing.

C.2.1 Variance

Variance is the expectation of the squared deviation of a random variable from its mean and in reference to a population, and it is expressed as

$$\sigma^2 = \frac{\Sigma(x_i - \mu)^2}{N} \tag{C2}$$

where N is the size of the population, μ is the population mean, and x_i are the dispersed values of data. Values are squared in order to neutralize the effects of negative and positive deviations in estimating the dispersion.

The variance of a sample is similar to the form:

$$s^2 = \frac{\Sigma(x_i - \bar{x})^2}{n-1} \tag{C3}$$

where n is the size of the sample, $\bar{x}$ is the sample mean, and x_i are the dispersed values of data. The sample population in the denominator is reduced by one in order to achieve an unbiased estimate (Hogg & Tanis, 1988). The square root of σ^2 or of s^2 is called the *standard deviation* and is a fundamental descriptor of dispersion.

C.2.2 Range

Range is taken literally in statistics. It is the difference between the smallest value and the largest value of a set of data and, as such, is the simplest measure of dispersion. Until relatively recently, range was commonly used in Shewhart control charts because of its simplicity, but it had to be used with

approximation factors to improve its accuracy. Today, with handheld calcula-
tors, the variance of a data set is easy to determine and most control charting
now uses the more accurate estimate of the standard deviation.

C.3   Distributions

Almost all, if not all, natural events have a random component. Mathematicians,
scientists, engineers, and sociologists, among others, study randomness and
build probability models to help in predicting events. Nature is under no obligation to follow our models, but often the models are quite good at describing
the random events of concern. Weather prediction is a popular example of
the use of probability models in prediction. Because these probability models are
designed to show how an event relates to the probability of its occurrence, the
measure of the event is distributed along the x-axis and its probability of occurrence is distributed along the y-axis. For this reason, a probability model is
called a "probability distribution," or "distribution" for short.

C.3.1   Continuous Distributions

There are at least hundreds of distributions, but the most well-known one is the
normal curve, often called the bell curve by sociologists and the Gaussian curve
by engineers. The normal distribution is shown in Figure C.1 and is generated
by the equation

$$f(x) = \frac{1}{\sigma\sqrt{2\pi}} \exp\left[-\frac{(x-\mu)^2}{2\sigma^2}\right] \tag{C4}$$

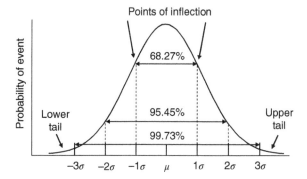

Figure C.1 The normal distribution.

where μ and σ are the distribution mean and standard deviation, respectively. The bracketed expression in this equation is the exponent of the natural base, e, and written larger here to more clearly show the parameters in the exponential term.

Normal distributions are completely described by their mean and variance and are often denoted as $N(\mu, \sigma^2)$. A special case of the normal distribution, called the *standard normal distribution*, occurs when $\mu = 0$ and $\sigma = 1$. These values assure that the total area under the curve is unity and is particularly useful in a number of statistical applications. The standard normal distribution is denoted $N(0,1)$ and its values are tabulated in many mathematical and engineering textbooks.

The normal distribution is mound-shaped and symmetrical, which means that the probability of being greater than the mean value is equal to that of being less. The crest of the form is a positive (convex) curve. Conversely, as we descend in either direction, we arrive at a negative (concave) curve. Where these two curves meet is called *a point of inflection* and offers a natural index for measuring deviations from the mean.

In Figure C.1, the x-axis is indexed in units of σ, the standard deviation. Thus, an event, x, can be referenced to the standard deviation of its distribution. The y-axis represents the probability of an event, x, with the highest probability being the mean value, and divergent values being symmetrically displaced about the mean.

All normal curves have the same symmetrical form and the width of the form depends on the standard deviation of the population distributed. One standard deviation is that point on the x-axis that coincides with a plumb line dropped to the x-axis from either point of inflection on the curve below and above the mean—always.

Thus, it is the magnitude of the standard deviation that determines the width of the normal curve. The larger the deviation from the mean, the wider the curve. As variation from target value is undesirable in product or service, a relatively thin curve would be preferable to a wide curve. Ideally, variation would be zero and the key characteristics of all products would have a mean value equal to the target value. From a system's engineering perspective, the wider the distribution, the less capable is the process that makes the unit because a wide curve represents increasing probability of unit values approaching or exceeding the tolerance limits of acceptable design.

As a normal distribution is continuous, it is meaningless to speak of the probability of an event at a specific point in time—there are an infinite number of points on the line of real numbers. Rather, tables of probability in the normal distribution indicate the probability of an event occurring within a range of values that define an area under the normal curve, the range usually specified in terms of Sigma. For example, Figure C.1 says that the probability of an event occurring between $\pm 1\sigma$ is 0.6827, which is the area under the curve that lies between the $\pm 1\sigma$ indices. Thus, 68.27% of the area within the curve is in this range. Using the same reasoning, the probability of an event occurring between $\pm 3\sigma$ is 0.9973.

An event that occurs beyond $\pm 3\sigma$ is said to occur in the tails of the distribution. The probability of an event occurring in the tails is $1 - 0.9973 = 0.0027$, a very low probability and by convention, a rare event.

It is common in industry to keep track of three standard deviations from the mean value in either direction. According to Walter Shewhart (1931), there are economic reasons for doing so. The tails of the distribution take on great importance in deciding whether a process is stable and in estimating the probability that a sample value in the tails derives from that distribution or from some external disturbance.

Sigma can be expressed in any units—whatever it is you are measuring. If the event is being measured in, say pounds, then the mean value and the standard deviation will be in pounds. If the event is being measured relative to time, then Sigma is expressed in time. Time helps to give a feel for the normal distribution, for suppose that a random variable is known to be just as likely to occur early as to occur late. Then its probability is mound-shaped and might well be modeled with a normal distribution.

Not all random variation behaves symmetrically. Some distributions are therefore skewed, as for example the distribution in Figure C.2a. A skewed distribution with its long tail to the right is said to be skewed right. In terms of probability, this type of skew occurs when a random event occurs sooner rather than later. Delivery times tend to be skewed right. Sometimes it seems that contractors you hire are skewed left—they tend to show up later rather than sooner!

C.3.2 Discrete Distributions

Basically, there are two kinds of measured data: variable and attribute. Variable data assume values that are continuous—they extend over an infinite continuum. Consider length for example. A measurement can be 33, 33.333 ft, and so on, limited only by the precision of the measuring device. Attribute data, also known as count data, are discrete because the data are in whole numbers, or their ratios. Five errors per page is a countable number. The ratio of 53 nonconformities in 497 tests = 0.1066398. This ratio is a failure rate. Although it appears to be a continuum, it is computed from count data and therefore it is considered an attribute. Why? Because the number 0.1066398 is not the measurement. It is a computed value. The measurement is 53 out of 497, which are count data.

Measured data and count data have different distributions. Measured data have continuous distributions. The distributions are not necessarily normal but are often mound-shaped. Two rather common continuous distributions that are not mound-shaped are the uniform and the exponential. A distribution that has no mode because all values are equal in amplitude is called

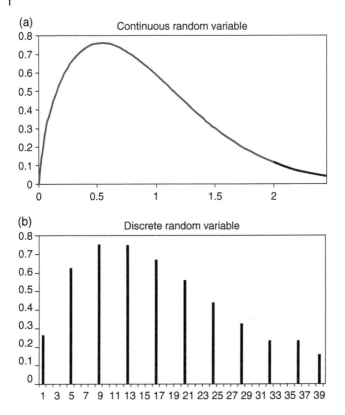

Figure C.2 (a) Continuous and (b) discrete distributions skewed right.

uniform. The exponential distribution is one that decays exponentially from some initial peak value.

Count data have discrete distributions as shown in Figure C.2b. A useful random variable from a forensic systems viewpoint is the fraction defective of a population, which has a binomial distribution. The binomial distribution is appropriate for discrete events that are mutually independent, have a yes or no outcome, and have a constant probability. This condition describes the attribute samples of a stable process. Suppose that a manufacturer produces integrated circuit cards, which are then tested in a go–no-go test at the end of the production line. The ratio of cards that fail the test to the daily total production is distributed binomially and is a measure of the effectiveness of the process.

Discrete distributions can be formed in two ways. The first is to simply measure a discrete random variable, say the monthly sales of an automobile. The other is to measure a continuous random variable in discrete time, say the thickness of plastic sheeting. If in reference to increments of time, either type of measurement is called a time series. Analysis of time series is a major part of forensic systems investigation, as described in Chapter 15.

It is useful to compare the form of the two distributions in Figure C.2. They have similar shape—a mound-shaped distribution skewed right, and indeed, the same statistical methods can be used on both. This approach is true in all cases, subject to certain conditions, and some of the analyses are described in detail in Appendices D and E.

C.4 Tests of Hypotheses

A test of hypothesis is a method of statistical inference concerned not so much with estimating the value of a parameter as with testing a hypothesis about it. Usually, the hypothesis is concerned with whether the parametric value has changed from some expected value. If you are a designer attempting to improve a process, you will want to test whether the process redesign improved its performance. Your hypothesis would then be posed in the form of a question requiring a yes or no answer. In forensic systems investigation, we are often more concerned with whether a change caused a degradation of performance.

The answer is approached with two hypotheses: the null hypothesis, called H_0, and the alternative hypothesis, H_a. The null hypothesis is NO! there is no change to the parameter and thus no improvement. The process remains as it was. The alternative hypothesis is YES, the parameter has changed significantly, implying an improved process. Of course, the same reasoning applies for testing degradation as well as for improvement. If subsequent testing indicates that there has been no change, then H_0 is true and H_a is false. Conversely, if testing indicates that the process had a statistically significant change, then H_0 is false and H_a is true.

By convention, the null hypothesis is assumed true going into the test. It is usually the Devil's Advocate position, posed in such a way as to say that whatever you are trying to prove is wrong. The reason for this convention is that by beginning with the assumption that you are wrong keeps the game honest—you must prove that you are correct.

C.4.1 Estimating Parametric Change

The hypothesis test is done by sampling processes that are discussed in other appendices. At this point, it should be noted that statistics is the study of uncertainty and can never establish the truth beyond the shadow of doubt. Hence the term "statistically significant," which refers to being able to distinguish that which is highly probable from that which is less probable.

Suppose that a certain process is believed to have a nonconformity rate no greater than 3%. However, a production engineer studying current data believes that the process has shifted with time and that the rate has increased. The engineer wants to test the claim and chooses a sample size sufficient to ensure that a normal approximation is appropriate for the binomial distribution. The sample size is sufficient if both np and nq are greater or equal to 5, where n is the sample size, p is the failure rate, and q is unity minus the failure rate, that is, $q = 1 - p$.

Let $\hat{p}$ represent the sample nonconformity rate of the process and let p_0 represent the most probable nonconformity rate. The null hypothesis is that the claim is correct and that the engineer must prove differently. Therefore, the null hypothesis is H_0: $p_0 \leq 0.03$. The alternate hypothesis is that the process has shifted and the process nonconformity rate has increased beyond the claim. The alternate hypothesis is then H_a: $p_0 > 0.03$. The argument is shown in Figure C.3. The engineer wants a test at the 95% level of confidence in the test results and therefore locates the rate of 0.03 at the x-axis value corresponding to 95% of the area of the normal $N(0,1)$ curve. The remainder of the distribution above that value is defined as the α area or rejection region because the null hypothesis will be rejected if the resulting analysis shows $p_0 > 0.03$.

Tests of hypotheses are often called z-tests, where the parameter "z" is a test statistic associated with the area under the curve of a normal standard distribution. Therefore, "z" corresponds to values of σ over the area of $N(0,1)$. When $z = 1$, the area under the curve from $-\infty$ to z is 84.13% of the total area. Hence, the probability of a normally distributed event occurring between $-\infty$ and $z = 1$ is 0.8413. When $z = 1.645$, the area under the curve from $-\infty$ to z is 95% of the total area. The probability of a normally distributed event occurring between $-\infty$ and $z = 1.645$ is 0.95. Therefore, the area beyond this value

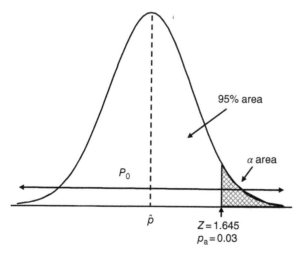

Figure C.3 A hypothesis of process nonconformity.

of z is 5% of the total. This area is in the upper tail of the distribution, the area of very low probability, or rare events.

To begin, we must decide on a statistically significant level for the test, that is, the level to which a measured variable can be distinguished from noise. We have selected 5%. Then in Figure C.3, if p_0 is located anywhere in the 95% area of the distribution less than 0.03, the null hypothesis will be accepted. The area above 0.03, called the α area, is the *level of significance* and represents the probability of mistakenly rejecting the null hypothesis when it is true.

A table of the normal distribution shows that 95% of the area under the curve occurs at a z-value of 1.645. If a sample provides a test statistic, $z > 1.645$, then the process is operating in the α area and H_0 should be rejected, as shown by Equation C5:

$$|z| = \frac{|\hat{p} - p_a|}{\sqrt{\hat{p}\hat{q}/n}} > 1.645 \tag{C5}$$

where $\hat{p}$ is the ratio of the number of defects in the sample to its size, n, and $\hat{q} = 1 - \hat{p}$.

The value, p_a, is a reference probability, in this example the claim of 0.03 that is challenged by the production engineer or in general, possibly an acceptable deviation rate. In any case, from a forensics point of view Equation C5 is not very efficient because the hypotheses have an infinite number of values. We can determine whether p_0 is or is not in the rejection region, but we do not know its limit, which is needed in order to be compared to an acceptable deviation rate. Equation C5 is used simply to explain the notion of hypothesis testing. A much better approach to determine process nonconformity limits is shown later in this appendix in the section on ordered statistics.

The level of significance puts a bound on the error in judgment that can occur in estimating an uncertain outcome. In the case just described, we can be 95% confident in our decision but can never be completely confident. If the test leads to the conclusion that H_0 is false when in fact it is true, then we have committed a type I error. If the test leads to a conclusion that H_a is true when it is not, then we have committed a type II error.

What is the nature of these errors? They are improbable. In the case just described, there is only a very small probability that test results will indicate H_0 is false when it is true—less than a 5% chance. Conversely, there is only a small probability that test results will show H_a to be true when it is false, although this is not apparent here. The evidence for an error in H_a is shown later in Appendix D, in the discussion on β errors. Table C.1 provides a matrix of the decision outcomes of a test of hypothesis.

A type I error occurs when the auditor decides, on the basis of the sample, that the process has changed when it has not. H_0 is true and the auditor declares it false. A type II error occurs when the auditor decides, on the basis of the

Table C.1 Errors in a test of hypothesis.

	True system state	
Conclusions	H_0 true	H_a true
H_0 true	Correct decision	Type II error
H_a true	Type I error	Correct decision

sample, that the process has not changed when it has. H_a is true and the auditor declares it false. The culprit in each case is not the sample, which is nothing more than a set of numbers. The culprit is the auditor, who has unwittingly misinterpreted them.

C.4.2 Confidence Level

Two terms of frequent use in statistics were mentioned earlier: *level of significance* and *confidence*. The first refers to the statistical significance of the test—the level at which the cause of a measured variation can be distinguished from noise. Thus, $\alpha = 0.05$ means we can distinguish designed effect from noise down to 5%. Hence, α represents the level of significance and can be selected according to need. It follows that the confidence that we can have in the measurement is related and is defined by a *confidence coefficient*, Cc:

$$Cc = 1 - \alpha. \tag{C6}$$

When the confidence coefficient is expressed as a percentage, it is called the *confidence level*. Thus, if $\alpha = 0.05$, then $Cc = 0.95$ and we have a confidence level of 95% that the hypothesis test will yield a valid estimate.

C.5 Ordered Statistics

Let X be a random sample of values $x_1, x_2, \ldots, x_n$. If the values are ranked in increasing or decreasing magnitude, they are called *ordered statistics* and lend themselves to efficient means of determining the extreme values of X. This approach is particularly useful in determining, say, the highest deviation rate likely from a distribution represented by the sample. For example, a sample size of 67 containing 2 nonconformities suggests a deviation rate of about 3%. However, we know that every sample taken from the process would be different. Short of taking hundreds of samples for a best estimate of the true deviation rate, it is useful to determine the highest likely deviation rate from the sample of 2/67. You can find this answer with Equation C5, but it could take many iterations.

A more efficient approach is to model the sampled data as from a Beta distribution, which is particularly useful where not much data are available. The Beta distribution can take on a variety of forms depending on the values chosen for its shaping parameters. Then the inverse transform of the Beta distribution can generate the order statistics necessary to select the largest value at any given probability (Law & Kelton, 1991). The efficiency of this approach comes from available, easy-to-use software.

For example, the EXCEL application, BETAINV (Beta inverse), requires as inputs only the desired confidence coefficient, the deviation count, and the sample size. Suppose that in a sample size, n, the deviation count, x, is 2. Let $n = 83$ and $x = 2$. The sample deviation rate is thus 2/83, or 2.4%. Given this estimate, what is the highest probable deviation rate with a confidence of 0.95? BETAINV will answer this question with just three inputs: (i) the desired confidence, 0.95; (ii) the first shaping parameter, $x + 1 = 3$; and (iii) the second shaping parameter, $n - x = 81$. The result from BETAINV is the value 0.074, or 7.4%, as an upper deviation rate for this process. In sum, we can be 95% confident that a system that yields a sample of two nonconformities from a sample size of 83 has a deviation rate no greater than 7.4%. This claim could very well stand up in court irrespective of claims to the contrary.

(Warning to users of BETAINV: Version EXCEL 2003 uses the titles *Alpha* and *Beta* for the shaping parameters. I do not use these titles here because many readers will have other versions of EXCEL and because these titles are somewhat unfortunate as they have nothing to do with auditing errors traditionally called Alpha and Beta. In this book, they are simply called the first and second shaping parameters.)

References

Hogg, R. V. and Tanis, E. A. (1988). *Probability and Statistical Inference*. New York: McMillan, p. 327.

Law, A. M. and Kelton, W. D. (1991). *Simulation Modeling and Analysis*. 2nd edition, New York: McGraw-Hill, pp. 473–474.

McClave, J. T. and Dietrich F. H. (1988). *Statistics*. San Francisco, CA: Dellen, p. 32.

Russell, J. P. (1997a). *The Quality Audit Handbook*. Milwaukee, WI: Quality Press, p. 176.

Russel (1997b), p. 177.

Shewhart, W. A. (1931). *Economic Control of Quality of Manufactured Product*. Princeton, NJ: Van Nostrand, pp. 26–34.

Appendix D

Sampling of Internal Control Systems

In the event of litigation concerning a performer's operations, forensic systems analysis of the evidence in discovery may be necessary. Invariably, the issue will concern the performer's internal control system, which is the responsible function for the effectiveness of all operational processes. Samples of the evidence may be taken and to make valid inferences from the sample, the analysis should follow well-defined sampling plans that are recognized throughout the world.

There are two kinds of sampling plans: statistical and nonstatistical. These plans are discussed in detail in other appendices, but a few brief comments are appropriate here. Statistical sampling applies the laws of probability to the sampling scheme to measure the sufficiency of the evidence gained and to evaluate the results. Sometimes it is impractical or difficult to meet the requirements of

Forensic Systems Engineering: Evaluating Operations by Discovery,
First Edition. William A. Stimson.
© 2018 John Wiley & Sons, Inc. Published 2018 by John Wiley & Sons, Inc.

statistical inference. For example, randomness and sample size may be difficult to achieve in what may prove to be a partial audit. A large number of the process controls may not be sampled at all. Therefore, forensic systems analysts may find a nonstatistical sampling plan sufficiently effective.

A sampling plan is *nonstatistical* when it fails to meet the criteria required of a *statistical* sampling plan. Nonstatistical sampling plans rely primarily on subjective judgment derived from experience, prior knowledge, and current information in order to determine sample size and to evaluate results. A well-designed nonstatistical sampling plan can provide results just as effective as those from a statistical sampling plan, except that it cannot measure sampling risk. Therefore, the choice of a statistical or nonstatistical sampling plan depends upon the trade-off of costs and benefits relative to sampling risks.

However, in implementing a nonstatistical sampling plan, procedures should be similar to those of statistical plans in order to safeguard the integrity of the findings. The analyst must understand the vocabulary and principles of statistical sampling. In this chapter, we shall discuss some basic statistical notions appropriate to all sampling techniques and delay the specifics of statistical and nonstatistical approaches to additional appendices.

Sampling, as most other technical fields, creates its own vocabulary. In the pages that follow, the reader will learn terms that come from the fields of sampling, the financial world, and manufacturing, and although the terms may have different names, many of them have the same meaning. Because forensic systems analysts read documents from these various fields, it is incumbent on them to be familiar with each term and understand its equivalency, where it exists. For example, the terms Beta risk, Beta error, Type II error, consumer risk, and probability of acceptance all describe similar events.

D.1 Populations

Sampling is the process of taking samples from a *parent population*, and based upon the statistics of the samples, making an estimate of properties of the population. *Population* is a statistical term for a set of objects under study. It could be what we normally think of as a population—say the population of cod off the coast of Iceland, or the population of people in Muncie, Indiana. But "population" can also refer to any large set—the professional football games played between 1960 and 1990, the crop of vinifera grapes in Monterey county in 1968 through 1988, the customers that use a local bank on Saturdays, or the week's production of integrated circuit cards in a factory.

Forensic systems analysts are interested in the population of controls, and specifically in the mean value, variance, and distribution of their output. The mean and variance of the population are called its *parameters*. The samples provide estimates of the mean and variance and are called *statistics*. The sampled data can also indicate the distribution of the parent population. Knowledge of the distribution of a population enables an estimate of whether a sample is representative of that distribution or of some other. This is important to the analyst because it aids in the estimate of the effectiveness of the controls. A stable process provides key characteristics with a distribution. If this distribution has been identified and products from the process do not fit the distribution, then clearly the process has shifted and is functioning in a new, perhaps unknown distribution.

D.1.1 Sample Populations

The task in forensic systems analysis is to verify the existence and effectiveness of the controls of the processes in litigation. If the set of controls under consideration represent a sample from discovery, then they are considered the *sample population*. In this context, the controls are those policies, procedures, documents, and actions that the company employs to ensure compliance of their operations to contract and legal requirements and conformance to good business practices. A nonconformity in the sampled population is often called a *deviation* because it deviates from the specifications.

Writing about financial systems, Apostolou & Alleman (1991a) describe it this way: An auditor tests controls to see if they are functioning properly. The auditor measures compliance with the control procedures in terms of deviation rates. Compliance in a given test means the control works—control is present. Noncompliance means that the control does not work—control is absent. A deviation exists. The judgment of compliance is made by comparing an estimate of the process deviation rate made from sampling, to an acceptable deviation rate determined during the formulation of the sampling plan.

This sounds similar to an acceptance sampling scheme used in manufacturing, as well it should. Acceptance sampling is similar in kind to a financial audit. For example, certified public accountants (CPAs) conduct audits that are required by law, but they may use sampling because of the sheer scope of their audits. Noncompliance may be against the law and carry severe penalties, so a CPA must use correct sampling techniques and much of the literature on audit sampling derives from this profession. This is all to the benefit of forensic systems analysts, who are also subject to legal scrutiny and who are free to use the same techniques.

The objective of sampling is to make inferences about a parent population based on the statistics of a sample population. The inherent variation of a stable parent population is called system noise. The variation in the sample population will be an estimate of system noise. Because of the nature of randomness, each sample will yield a different value as an estimate, and as the number of samples increases, the average value of sample noise will more closely approximate the value of the system noise.

It is possible to introduce added variation into the measurement, caused by the measuring or audit technique, which is not itself part of system noise but will appear to be so. Great care must be taken in the design of a sampling plan to avoid introducing noise. For example, in the practice of forensic analysis, "noise" would be introduced by the improper assortment of evidence or by the comparison of nonhomogeneous characteristics.

D.1.2 Population Size

How large must the audited population be to be considered "very large?" Experts offer a variety of views on the matter—Apostolou & Alleman (1991b) state that a population greater than 500 would be sufficient. Other sources say that several thousands would be necessary. Grant & Leavenworth (1988a) state in reference to attribute sampling that as long as the parent population is at least 10 times the sample population ($n/N < 0.1$), the binomial distribution can be used. Mills (1989a) echoes this ratio, adding that in auditing generally the parent will be much larger than the sample population and simplifying statistics can be used. Guy (1981a) uses the proportion of populations where $N = 1000$, $n > 20$, and $n/N < 0.1$. These numbers are similar to those expected of an internal audit. Therefore, if attributes are measured, then it is reasonable to use the more simple binomial distribution in audit sampling in lieu of the hypergeometric distribution.

D.1.3 Homogeneity

In assembling the evidence in some desired order, the data should be grouped on the basis of homogeneity. The members of a homogeneous population are pretty much alike, relative to the purpose of the inquiry, which allows you to make valid statistical inferences about it. As an easy example, suppose that you want to evaluate oranges on the basis of flavor. Then all the members of the population must be oranges. If you want to evaluate a welding process, then the sample must contain only those items having been welded by that process. An unsigned blueprint, although important to another evaluation, would not be a member of the population of welded units. These three examples of determining homogeneity are clear; other cases might be more difficult to discern. For example, suppose a single machine produces nonconformities of two different kinds, say dimensional and scarring. Whether they are homogeneous nonconformities depends on how the question is framed.

An inquiry into operations will include documentation, management, resources, operations, supplier control, process stability, and verification and validation of product and process. Each activity will have its own metrics. It may be necessary to divide the activities into similar groups.

The division of a nonhomogeneous population into homogeneous groups for purposes of sampling is called *blocking* or *stratified sampling*. Then sampling is taken from the groups and the average and variance of each activity or process is determined. Based upon the statistics of each group, an estimate can be made of the overall picture of control system effectiveness.

Generally, the population of evidence in discovery is not homogeneous. Many diverse activities are examined against different requirements. However, you can cleverly define the controls so to be homogeneous, in creating a total population sufficient for statistical analysis of the overall system of operations, yet by still allowing for assessment of the performance of each process. This can be done by employing a technique called attribute sampling in which controls and decisions are treated as attributes.

D.2 Attribute Sampling

In everyday language, an *attribute* is a synonym for a characteristic or a property of something or someone. We might say, for example, that George Washington had the attributes of courage and resoluteness, or that mica has the attribute of translucence. In forensic systems engineering, however, an attribute refers to a binary property. For example, a control either has a defect or it does not, it is either in conformance to requirements or it is not.

In general terms, there are two measures of a parameter in product or service. If the parameter can be measured dimensionally, for example if it has thickness or length or weight, then it is called a variable and its values are defined over a continuous range. If the parameter is measured as simply either acceptable or not acceptable, then it is called an attribute and has only one of two values, 0 or 1.

As an example, consider the key characteristic of a product. It may be a count or it may be a dimension. In either case the characteristic either conforms to specifications or it does not. If the output complies with specifications, the control works—control is present. Noncompliance means that the control did not work—control is absent. The presence or absence of a control is an attribute. Therefore, attribute sampling can be used by forensic systems analysts to arrive at a numerical evaluation of the effectiveness of controls of a given process.

In financial auditing, attribute sampling plans represent the most common statistical application that is used to test the effectiveness of controls and to determine the rate of compliance with established criteria. The results of these plans provide a statistical basis for the auditor to conclude whether the controls are functioning as intended, reflecting either compliance or noncompliance, hence a binary (yes or no) decision.

Attribute sampling offers an additional advantage to forensic systems analysis because in this type of sampling, many of the engineering terms and the financial terms become equivalent, as do their procedures. Thus, the integrity of forensic systems analysis can be identified with that of financial audit procedures so firmly established in law. The equivalency of these terms is shown in the sections that follow.

D.2.1 Acceptable Deviation Rate

Variation exists in any scheme of production and service. Therefore, an acceptable rate of nonconforming units must be defined, similar to an acceptance number used in sampling. In finance this rate is the *acceptable deviation rate* (ADR). The equivalent term in manufacturing is the *acceptable performance level* (APL). Recognizing that no sampling plan is perfect and that there will always be undetected nonconforming units (escapes), it can be shown that a well-conceived sampling plan with an APL performs better than sampling the entire population (Grant & Leavenworth, 1988b).

The notion of an "acceptable" deviation or performance rate is controversial. In the early history of quality control, the idea was opposed on principle (Grant & Leavenworth, 1988c). Why should a customer pay for a lot knowing that it contained defective product? Regarding the evaluation of a process, Mills (1989b) states that any nonconformity in a sample requires corrective action; therefore, the auditor is looking for error-free performance and a zero acceptance number. According to Mills, any failure is intolerable and the APL is specified simply to determine a sample size.

In defense of the APL, it is argued that a sampling plan with an APL has superior operating characteristics than those of a sampling plan with an acceptance number of zero (Grant & Leavenworth, 1988d). It is generally recognized today that there will always be a process deviation rate because no production system is perfect, nor is any sampling method perfect. Therefore, the APL is accepted as an economic solution to this reality.

An APL measure is a factor in the design of a quality system, but Forensics is the study of that which has already happened. The design phase is over and the system is operating. The forensic systems analyst is concerned with a binary issue: whether or not a control is in conformity to requirements, or whether it deviates from them. Hence, ADR is the more appropriate term for forensic use and we shall adopt the term "acceptable deviation rate" rather than "acceptable performance level."

If in your study you find zero nonconformities, it does not mean that there are none, it simply means that you did not find them. If you find one or several nonconformities, it may not mean that the control is ineffective. It means that the company has decided that a certain amount of variation is acceptable in any epoch in the development of its products and systems. If this decision is acceptable to the customer, it must be acceptable to the forensic systems analyst. It is incumbent on forensic analysts to consider the acceptable deviation rate in their analyses and assessments.

D.2.2   System Deviation Rate

Every system varies randomly in the quality of what it does, including variation of its key characteristics. If the system is stable, the variation will be acceptable for the most part and the nonconforming units of output will be relatively rare. If the nonconforming units are counted as a proportion of the total, then the ratio of nonconforming units to the total population counted is called the system deviation rate, or SDR. This reasoning applies also to a control process.

We usually do not know the system deviation rate, but derive an *estimated system deviation rate* from the control samples. We really need to make *two* estimates—a pretest estimate and a posttest estimate. The pretest estimate is made to determine the required sample size needed to measure the effectiveness of a control. The size depends on the difference between the acceptable and system deviation rates. The posttest estimate, derived from the test itself, will measure the control effectiveness.

The pretest estimate can be based on observation, inquiry, or experience, but some knowledge of system capability must be available. This condition is not at all unusual; Box et al. (1978a) stress repeatedly that statistical techniques are most effective when combined with appropriate subject matter knowledge.

Thus, we begin with an *expected system deviation rate* that can serve to determine the sample size, then take the sample and obtain the estimate of the SDR. The auditor compares this system deviation rate to the acceptable deviation rate. If the SDR is lower than the acceptable rate, subject to appropriate confidence level and sampling risk, then the control procedures are deemed effective. Otherwise, they are deemed ineffective. The procedures for determining the sample size with an expected SDR are explained in a later section of this appendix.

D.2.3   Controls

We have established that the presence or absence of a control is an attribute. In this sense, "absence" is taken both literally and figuratively, referring both to when there is no control where there should be one, and to an existing control that does not control. A control either works or it does not, thereby verifying the conformance of the process.

For example, consider again the receipt inspection process described in Chapter 17. The procedure defines the process with six controls:

1) Receiving. The parts are received at a point designated as the receiving station.
2) Holding. An operator labels the parts "not inspected," and then puts them in a quarantine area until they can be tested.
3) Inspection. An inspector inspects or tests the parts for conformance to the purchase order, according to a specified acceptance scheme.

4) Labeling. The inspector labels the good parts as "Inspected OK," and labels the bad parts as "Inspected and unacceptable: return to vendor."
5) Recording. The inspector records the measurement results as a quality record, sending a copy to purchasing.
6) Traffic. A forklift operator moves the good parts to inventory or to production and moves the rejected parts to a specified shipping dock.

Suppose the evidence indicates that one of the controls is not being followed, but the rest are good. You now have a judgment call. You can either assess receipt inspection as nonconforming or assess the process as having a deviation rate of 16.7% (one defect in six controls) and compare it to the acceptable deviation rate if one exists. In this case, the rate is probably too large even for a process in development and receipt inspection is not usually regarded as a process in development.

Your assessment is a go/no-go decision no matter the nature of the control. Mills (1989c) asserts that by defining all control assessments as yes or no decisions means that the entire control population can be treated as homogeneous. This provides a sufficiently large population, which as we shall see in the discussion on statistical sampling plans is one of the considerations necessary to valid statistical inference.

D.3 Sampling Risks

D.3.1 Control Risk

For systems analysts, there is something of value in emulating financial auditing. Control risk has been defined under *International Standards of Auditing* as the risk that a misstatement that could occur in a financial statement and that could be material, either individually or when aggregated with other misstatements, will not be prevented, or detected and corrected by internal control (ISA 200, 2010). If we replace the word "misstatement" with "misfeasance," then the definition used in Chapter 17 corresponds: A control risk is the risk that a given control structure may fail to do its job. This issue is addressed in detail in Chapter 14. Relative to sampling, control risk can be regarded as the risk of a false assessment of control effectiveness due to sampling errors.

D.3.2 Consumer and Producer Risks

In manufacturing operations, there are two business risks that can occur: *consumer risk* and *producer risk*. As no acceptance sampling plan is perfect, it is possible to accept and ship a lot that contains nonconforming units. The probability of accepting such a lot is called a consumer risk for obvious reasons. There are bad units in the lot and some customers will purchase them. On the other hand, it is also possible to reject a lot with an acceptable proportion of units that conform to requirements. This is called a producer risk because it adds unnecessarily to the cost of production, condemning for rework or further inspection of a lot containing few defectives. These risks are unavoidable, but there are statistical techniques that can assess the inherent producer and consumer risks of any sampling plan. Consumer and producer risks are inversely related, and management must make a decision on an acceptable level of these risks. Hypothesis testing, discussed earlier in Appendix C, can be used as the process for selecting this acceptable level.

D.3.3 Alpha and Beta Errors

The best sampling plan in the world can provide only an estimate of system performance. The statistical deviation rate is only an estimate of the system's true deviation rate, and the statistical variance is only an estimate of system variance. Yet, the analyst must make a decision about the system based on these estimates. There is a risk of being wrong, one way or the other and the two ways of being wrong are known as *alpha* and *beta* errors.

The alpha error is that the sample will indicate that an effective control is ineffective. The beta error is that the sample will indicate that an ineffective control is effective. The two errors are also known respectively as Type I and Type II errors, or producer and consumer risks, or alpha and beta risks. The multiple naming of the same ideas can be vexing to initiates, but comes about because the concepts were originally approached in different professional fields: manufacturing, business, statistics, and mathematics.

The analyst makes an *alpha error* by deciding, on the basis of the sample, that the control is ineffective and there is a high risk that nonconforming units are going out. However, in reality the control is effective, with few nonconforming units. From this description, the alpha error is clearly a producer's risk, having rejected a control of acceptable quality. It is also akin to Type I error as it judges the status quo to be wrong when it is correct.

The analyst makes a *beta error* by deciding, on the basis of the sample, that the control is effective, with low risk of nonconforming units. However, in reality the control is ineffective, with an unacceptable level of nonconforming units. Thus, a beta error is a consumer risk because this unacceptable level will be delivered to the customer. It is also akin to Type II error as it judges the status quo to be correct when it is wrong. Table D.1 relates these concepts to the sampled data and to the control system.

According to Guy (1981b), the audit profession does not regard alpha and beta errors as equally bad. Suppose an analyst commits an alpha error, deciding that a good control is not effective. If the analyst stops there, then harm is done because a producer's risk is assigned. The producer must take corrective action where none is really needed. But usually an analyst does not stop there because

Table D.1 Alpha and beta errors related to decision analysis.

Decision indicated by sample	Control procedure is	
	Effective	**Ineffective**
Accept	Correct decision	Beta error
Reject	Alpha error	Correct decision

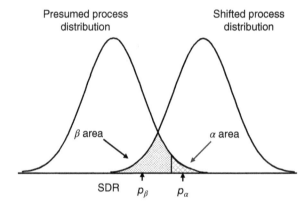

Presumed process distribution   Shifted process distribution

Figure D.1 The interaction of alpha and beta errors.

if a sample indicates the control should be rejected, the client will insist that the analyst look for more evidence and take more samples. Presumably, larger samples will be more informative. Hence, an Alpha error can lead to an ineffi-cient but effective audit.

If the auditor commits a beta error, deciding that a poor control is effective, then the decision will not be challenged, there is no motivation to continue, the review stops (at least with respect to the verification of that control), and harm is done because a consumer risk exists. Thus, a beta error leads to an ineffec-tive audit. For this reason, CPA auditors use beta error as a measure of audit effectiveness.

Engineers, too, focus on the beta error, defining the *power of the test* as $1 - \beta$, which is the probability that the null hypothesis will be rejected when the alter-nate hypothesis is true. The alpha error is thus minimized for any selected beta error of a sampling plan.

Figure D.1 may clarify what is going on between alpha and beta errors. The distribution on the left is the process presumed to be operating, based on design or on the claims of the process owner. The true mean value, SDR, may be unknown, so we take a sample from the process in order to estimate the SDR. Suppose that the sample has an average value, p_a, whose value is located in the upper tail of the presumed distribution, as shown in the figure. We cannot be sure that it derives from the process distribution; it could come from some other. For example the process performance may be operating differently than design, causing a shift in the process distribution, but unknown either to the analyst or to the owner.

Being in the upper tail of the presumed process distribution, p_α has a very low probability of deriving from that distribution; therefore, we may reject it. Indeed, by industrial convention we shall reject it and assume it derives from the distribution shifted to the right, to which it clearly has a higher probability of deriving. Yet it *could*, albeit with low probability, be a member of the presumed distribution and by rejecting it, we have made an alpha error.

But suppose that our sample is p_β, which is located well into the SDR space. On that basis we might accept it as deriving from the presumed disposition, as the probability is greater. But it *could* derive from the shifted distribution. If so and having accepted it, we have just made a beta error. Both of these descriptions apply to the other end of the spectrum also, that is, in the case when the system has shifted left.

As the image shows, a decrease in one of the risks brings about an increase in the other, so the management choice of a lower producer's error will at the same create a higher consumer's error, perhaps causing higher warranty reserves to cover future claims. The trade-off requires a well-reasoned balance and is always a management decision.

D.4   Sampling Analysis

By its nature, a sample from a population can provide only an estimate of a parameter of that population. Nevertheless, it is possible to determine the "goodness" of the estimate by statistical inference. In particular, statistical inference allows us to measure the errors in our sample. Inference is appropriate to a process that follows certain laws of probability with regard to randomness and independence. Let us review the laws essential to the argument.

D.4.1   Statistical Inference

Earlier, we reviewed the criteria required for a valid statistical sampling plan. The first criterion is that each member of the audited population be representative of the population. This is a requirement on homogeneity, to prevent comparing apples and oranges, as the old saying goes. For example, if a projected population of evidence contains both variable and attribute controls, you would be obliged to define two different populations. If the population contained subgroups of the same kind but with significantly different parameters, it would be nonhomogeneous. Homogeneity of a population can at times be achieved by how the population is defined, as was stated in an earlier citation that even diverse controls can be made homogeneous if they are defined as simple yes or no decisions.

The second criterion is on process stability. You cannot estimate the mean and variance of an unstable process because they do not exist, nor can you estimate the sampling risks. A stable process has a defined, if unknown, distribution and the whole purpose of sampling analysis is to assess the control effectiveness by estimating certain properties of that system distribution.

The third criterion is on randomness. As described by Box et al. (1978b), randomness is at the heart of much statistical analysis. A random sample is one in which each member of the audited population has an equal chance of being drawn. A member of a sample is a random variable if it was drawn from a known distribution. Introducing randomness into the sample improves our ability to use the sample for inference to the parent population.

The fourth criterion is that the sample members are mutually independent. This ensures that each member, a random variable, has the same distribution, so that the sample distribution, too, is stable. Randomness and independence let you make inferences about the parent population with a relatively small sample. You can use statistical inference without these properties, but then you would need a very large sample in order to make valid estimates of the parent population.

D.4.2 Sample Distributions

The probabilities associated with drawing samples from a finite population are described by a hypergeometric probability model. For example, acceptance sampling is the process of drawing samples from lots of manufactured product. Since the lots tend to be relatively small, say fewer than 500, the hypergeometric model is required to relate the probabilities of the sample to the probabilities of the lot. Unfortunately, this distribution is difficult to work with because it involves four variables: the population and defect rates of the lot, and the population and defect rates of the sample. If, however, the sampling is a Bernoulli process from a very large population, then a binomial probability model can be used, as the binomial involves only two variables, therefore simplifying calculations ensue.

A sequence of n trials is a Bernoulli process if the trials are independent, with each trial having only one of two possible outcomes, and with the probability of success remaining constant from trial to trial. The term "success" simply refers to a designated result—it could be the set of 1's or it could be the set of 0's, whichever we are tracking. The random variable, x, that denotes the number of successes in n trials has a binomial distribution. Assume that a very large population of controls, n, are audited as either in conformance or not, and that each control is selected randomly. Then the audit sequence is a Bernoulli process. Moreover, if x deviations are found, then the ratio, x/n, is a random variable with a binomial distribution.

D.4.3 Sample Size

If the size of the sample is sufficiently large, then the proportion nonconforming of the sample will mirror the proportion nonconforming of the lot. Most statistics textbooks say that the sample size can be found by solving equations relating to the sample, n, and the expected deviation rate, p (e.g., Equation D2). However, because stable control systems have quite low nonconformity rates, this procedure will often result in a large sample size that is impractical for forensic purposes. Moreover, the result will not have considered the consumer and producer risks that are essential to process analysis. One can include these risks by generating operating characteristic curves of the statistics in question.

Generating these curves require special software packages and a rather complicated procedure. Therefore, most auditors and analysts will use tables that provide sample sizes for various expected deviation rates and risks. Such tables are easily obtained in various books on sampling as it appears in the reference list of this appendix. In the example shown in the next section, a control was expected to have a deviation rate of 5% and an ADR of 8%. If we assume producer and consumer risks of 5% each, then the required sample size is at least 38 (Apostolou & Alleman, 1991c). This sample size is well within the range usually obtainable from discovery.

D.4.4 Estimating the SDR

The objective of sampling evidence is to measure the effectiveness of a set of controls by comparing the true system deviation rate of a control to its acceptable deviation rate. If the SDR is less than the ADR, the control is deemed effective; if not, then the control is deemed ineffective. However, we usually do not know the SDR, but must estimate it by using a sample deviation rate. To distinguish the two, the symbol $\hat{p}$ will represent the sample deviation rate. If we take a sample of size n and find x deviations, then

$$\hat{p} = \frac{x}{n}. \tag{D1}$$

In Equation D1, $\hat{p}$ is called the *maximum likelihood estimator* of the SDR. The number of deviations, x, may represent the number of product nonconformities from a given control, or it may represent the number of nonconforming controls in a given control system. In the first case, we are evaluating a control; in the second, we are evaluating a control system.

If the system is stable and we take, say, 30 samples of size n, each will have a different mean and variance. Given many samples, say, 1000, the average of the averages will approach the SDR arbitrarily closely. However, what can we do with just one sample? Given a stable process and attribute sampling, we can make from one sample a binomial point estimate of the system deviation, then determine from that point an interval estimate that is most likely to contain the SDR. This is called a *confidence interval* because it is the range of values that we are confident contains the SDR.

D.4.5 Confidence Interval

Sampling is an imperfect technique. Most of the desired results are only estimates of the true world. Sample size is based on an estimate. Control effectiveness is based on an estimate. The essential question is whether the estimate is close enough to the true value to achieve a correct decision. Fortunately, statistics also provides a way to determine a given confidence associated with the decision. In Appendix C, we discussed the confidence level that one could place on a test of hypothesis. A similar approach exists for determining a probable system deviation rate from a sample. We may not know the true distribution of a system deviation rate, but we can determine a *confidence interval* in which the SDR will be located. The usual confidence in this interval is of a magnitude between 90 and 99%.

The width of the confidence interval will vary according to the desired confidence. Suppose that we want to be 90% confident that the interval contains the SDR. Then the interval about $\hat{p}$ is

$$\hat{p} \pm z_{\alpha/2}\sqrt{\frac{\hat{p}(1-\hat{p})}{n}}. \tag{D2}$$

In Equation D2, z represents the number of standard deviations from the value $\hat{p}$ to which the interval will extend, that are associated with the probability of the desired confidence. The symbol, α, is the area of the distribution that extends beyond the interval—the area of rejection. It is divided by two because the distribution has two tails. So, for example, if you want to be 90% confident that the interval contains the SDR, you really want 45% in each direction above and below the mean. Thus, both tails are excluded from the interval. The tails of a distribution contain very low probabilities. The rejection does not mean that the SDR cannot be in the tails; rather, it means that if the SDR were in the tails, it would be a rare event.

Equation D2 also implies that we are using a normal distribution in lieu of a binomial distribution because of its well-defined characteristics. The binomial point estimate is unbiased and the binomial distribution approximated closely by the normal only if the ratio, x/n, is between the value of 0 and 1, and the closer to half way, the better. However, in stable industrial processes we would expect a rather low ratio, closer to 0 than to 0.5. This problem can be resolved in this case by a translation of the ratio closer to midpoint with a Wilson point estimate (Wheeler, 2011). The transformed Wilson point estimate is

$$\hat{p} = \frac{x + z^2/2}{n + z^2}, \tag{D3}$$

and the Wilson interval estimate is determined by

$$\hat{p} \pm z_{\alpha/2} \sqrt{\frac{\hat{p}(1 - \hat{p})}{n}} \tag{D4}$$

Often used values of $z_{\alpha/2}$ that correspond to various confidence intervals are shown in Table D.2. They and any other confidence level can be determined with a table of the normal distribution, listed as a column of standard deviations to the tenth and rows of hundredths. For example, 1.96 would correspond to 95% confidence (each tail being 2.5% of the total).

A visual concept of a point estimate with confidence interval is shown in Figure D.2. The alpha areas, or tails, are shown with respect to the 99% confidence interval. The tails would be greater and the width of the interval would be narrower for lower confidence intervals, as indicated in Figure D.2.

As an example, consider a count of 5 nonconforming units in a sample of 100 units. The estimated system deviation rate is then 5/100, or 5%, and assume that the assigned ADR is 8%. Let us take a look at a 90% confidence interval. Using Equations D3 and D4, and Table D.2, we arrive at a confidence interval extending from 0 to 12.7. (The lower limit is actually somewhat negative,

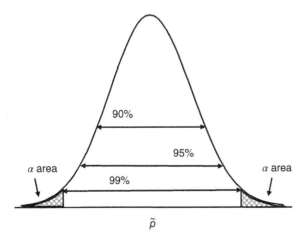

Figure D.2 An estimated deviation rate with confidence intervals.

Table D.2 Correspondence of standard
deviations to confidence level.

$z_{\alpha/2}$	Confidence (%)
1.645	90
1.96	95
2.576	99

theoretical, and irrelevant; negative deviation rates are undefined.) So we can be 90% confident that the SDR is located between 0% nonconforming and 12.7% nonconforming. Thus, the 5% sample deviation rate, being within this range, indicates that the sample comes from the SDR distribution and represents the expected process.

However, we must be aware of what we have established—the process seems to be operating as expected with a confidence of 90%. We have *not* established whether the SDR is less than the ADR. True, the upper limit of the interval, 12.7% is well beyond the ADR, but we do not know the probability of that 12.7%. It may be quite low. At this point, we might try the upper limit of the 5/100 ratio at 90% confidence with a BETAINV procedure. This search yields an upper limit of the sample distribution as 9.07%. This narrows the probable deviation width somewhat, but the upper limit still exceeds the ADR. Yet, we do not know the probability of the upper limit; it could be in the tail of the SDR distribution. Hence, the analyst must make a judgment call, understanding that grading the control as ineffective is justified but will be challenged.

References

Apostolou, B. and Alleman, F. (1991a). *Internal Audit Sampling.* Altamonte Springs, FL: The Institute of Internal Auditors, p. 51.

Apostolou and Alleman (1991b), p. 47.

Apostolou and Alleman (1991c), p. 119.

Box, G. E. P., Hunter, W. G., and Hunter, J. S. (1978a). *Statistics for Experimenters.* New York: John Wiley & Sons, Inc., p. 15.

Box et al. (1978b), p. 57.

Grant, E. L. and Leavenworth, R. S. (1988a). *Statistical Quality Control.* New York: McGraw-Hill, p. 220.

Grant and Leavenworth (1988b), p. 7.

Grant and Leavenworth (1988c), pp. 453–454.

Grant and Leavenworth (1988d), p. 403.

Guy, D. M. (1981a). *An Introduction to Statistical Sampling in Auditing.* New York: John Wiley & Sons, Inc., p. 57.

Guy (1981b), p. 133.

International Standard on Auditing 200 (2010). Handbook on Overall Objectives of the Independent Auditor and the Conduct of an Audit in Accordance with International Standards of Auditing, p. 77. http://www.ifac.org/system/files/downloads/a008-2010-iaasb-handbook-isa-200.pdf. Accessed August 25, 2016.

Mills, C. A. (1989a). *The Quality Audit.* New York: McGraw-Hill, p. 169.

Mills (1989b), p. 173.

Mills (1989c), p. 171.

Wheeler, D. J. (2011). "Estimating the Fraction Nonconforming." *Quality Digest.* https://www.qualitydigest.com/inside/quality-insider-column/051115-parts-million-problem.html. Accessed September 12, 2017.

Appendix E

Statistical Sampling Plans

Sampling is the process of selecting entities from a population of interest in such way that the characteristics of the sample may be generalized to the population from which they were chosen. In order that the sample reflects the characteristics of the total population, rigorous procedures must be followed, best assured by the use of a sampling plan.

A sampling plan is a detailed outline of which measurements will be taken at what times, on which material, in what manner, and by whom. A *statistical sampling plan* follows the laws of probability, allowing you to make valid inferences about a population from the statistics of the samples taken from it. It also lets you determine in advance the magnitude of the sampling errors.

Forensic Systems Engineering: Evaluating Operations by Discovery,
First Edition. William A. Stimson.
© 2018 John Wiley & Sons, Inc. Published 2018 by John Wiley & Sons, Inc.

There are many different statistical sampling plans, but we shall discuss just two, whose characteristics lend themselves to the partial audits taken from discovery. The first, *fixed-size attribute sampling*, can accommodate developing systems where system deviation may be relatively high and easily measured. The other is *stop-or-go sampling*, which can be used in stable systems with low deviation rates and hard-to-find errors.

E.1 Fixed-Size Attribute Sampling Plan

Sampling is efficient and practical when there is a trail of documentary evidence, or when the auditor can take the time to observe a process in action. Apostolou & Alleman (1991a) propose a seven-step strategy for a fixed-size attribute sampling plan:

1) Determine the objectives of the test
2) Define the attribute and deviation conditions
3) Define the population
4) Determine the method of sample selection method
5) Determine the sample size
6) Perform the procedures of the sampling plan
7) Evaluate the sample and express conclusions

E.1.1 Determine the Objectives

When we speak of testing operations, invariably we are speaking of testing its controls. The objective of the tests of controls is to assess the effectiveness of control procedures. One control might be a required sign-off signature or revision date on a document. Others might be a receipt inspection, feed rate on a machine, or completeness of a purchase order. The basic job of the analyst is to determine if the error rate of a system of controls is less than the acceptable level. As no control is perfect, there will always be output variation in any system. Therefore, given that a system of controls is not perfect, how good is it? Specifically, how does the measured system deviation rate (SDR) compare to the acceptable deviation rate (ADR)? If the measured deviation rate is lower than the acceptable rate, then the control procedures are deemed effective. Otherwise, they are deemed ineffective.

E.1.2 Define Attribute and Deviation Conditions

E.1.2.1 Acceptable Deviation Rate

By "attribute and deviation conditions," it is meant that the acceptable and SDRs of a control system are based on a count of its attributes. When performing tests of controls, you must specify the maximum number of deviations from the

prescribed control procedures that can be accepted and still regard the procedures as effective. This measure is also known as the tolerable deviation rate and
the acceptable error rate. To accommodate several disciplines, we use the term
acceptable deviation rate and explain how it might be used later in this chapter.

E.1.2.2 System Deviation Rate

The analyst usually does not know the SDR but must derive an *estimated SDR*
determined from the control samples. This estimation serves two purposes:
first, it is used to determine the sample size to be used in sampling from the
system of controls. Secondly, it provides the operational reference to be compared to the ADR.

E.1.3 Define the Population

The population of concern is all the items constituting a class of system controls. They must all have, nominally, the desired attribute and time period of
interest. As an example, if the attribute you are verifying is a properly authorized document change over the past 3 months, then any document changes
beyond this 90-day period do not belong to the population. A calibration
schedule change within the 90 days does not belong either, because you are not
looking for changes per se, you are looking for authorizations. The analyst
team must work together to arrive at a coherent sampling plan of nonredun-
dant, homogeneous controls.

Mills (1989a) suggests an effective strategy is to define all control assessments as yes or no decisions so that the entire control population can be treated
as homogeneous, albeit those decisions may relate to different technical points.
The homogeneity lies in the decision—to accept or reject a control. Whether
the control is a measurement, a document signature, observing a task, or
recording a result, the verification and validation must be posed to arrive at a
decision of acceptable or not acceptable. This convenience carries with it a
proviso on decision homogeneity—the block of controls should have similar
ADRs and sampling risks.

E.1.4 Determine the Method of Sample Selection

Once a strategy is decided upon, we are well on our way in developing a sampling plan. Now we must consider how the sampling itself will be done—the
taking of data. Some common methods of sampling are (i) random sampling;
(ii) systematic sampling; and (iii) haphazard sampling.

Valid statistical sampling requires randomization. Random sampling occurs
when each member of the population has an equal chance of being selected.
For example, suppose that you want to assess job orders for a given week. You
should not randomly choose from a set offered to you by the operations supervisor, but you must contrive to choose randomly from the entire set. They will

probably be stacked in some kind of order, so you must factor this bias in with your selection. For example, you might suppose that taking a few near the top, a few in the middle, and a few near the bottom are a random selection. They are not, because the job orders that are not near the top, middle, or bottom of the stack have no chance of being selected.

If you have defined the control population as all the job orders for the week, then each must be given an equal chance at being selected. You can, for example, assign numbers to the job orders, and then use a random number generator to help you in the selection. If this method causes you to end up with a random sample containing most of the job orders early in the week, so be it.

You might also choose the systematic sampling, which is choosing every nth member of the population, with the first being determined randomly. You need to know the population size and sample size. Suppose you are looking at job orders over the past year, and there are 5000 of them. You decide upon a sample size of 200. Then you begin at some random point and select every 25th member (5000/200 = 25).

You must avoid using haphazard sampling, which seems to be random but is not. This method consists of selecting samples without special reason and without conscious bias. At first glance this practice appears to be fair, but haphazard sampling can lead to samples that are not representative of the population. Because things are rarely arranged randomly, haphazard sampling will often include items that are not homogeneous. For example, suppose you are in the customer service office and you are asked to examine a batch of customer orders lying on the desk of one of the service agents. Will this batch reflect the effectiveness of control of that office? Not likely. If there were six agents in the office, it would be better to randomly select from the work of each one. In this way, each record has an equal chance of being selected. Haphazard sampling should not be used in statistical sampling, but it can be effective in nonstatistical sampling.

E.1.5 Determine the Sample Size

The advantage to statistical sampling is that sampling risks and confidence levels of the audit can be determined from the sampled data. One of the factors governing the validity of statistical inference is sample size. In order to determine an appropriate sample size, five factors come into play: (i) size of the total audit population; (ii) expected SDR; (iii) ADR; (iv) alpha error; and (v) beta error. A total population in excess of 5000 can be considered effectively infinite, which simplifies the statistics somewhat. However, that large a size is well beyond many audit populations.

According to Apostolou & Alleman (1991b), the effect of a population on sample size will not increase substantially as the population increases beyond 500. The knee-shaped curve of sample size versus population shown in Figure E.1

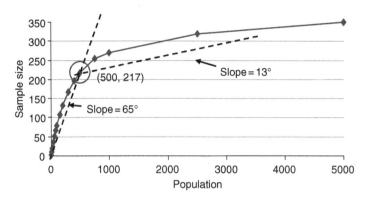

Figure E.1 Population size versus the required sample size.

supports the idea. Basically, two slopes define the response. The lower slope is 65° of rise to run. The breakpoint of the slope occurs at a population of about 500, where the required sample size is 217. The higher slope extending to an infinite population is much smaller, with a slope of only 13°. The curve is generated from equations on sample size (Hogg & Tanis, 1983).

The expected SDR can be estimated from prior records, if the controls and personnel have not changed, or from a pilot sample. Pilot samples are often integrated into an audit sample. For example, suppose your initial estimate of the SDR suggests a sample size of 124. You can select a pilot sample of 50, then test for accuracy. If the measured rate appears to support the initial estimate, you continue with the remaining 74 tests.

Alternatively, suppose your initial estimate of the SDR called for a sample size of 50, but the measured data indicate that a sample size of 124 should have been taken. Although your initial estimate was in error, the 50 tests are still valid and you need to take only 74 more samples to arrive at a valid estimate of the SDR. In this case, the initial 50 served as a pilot sample.

Many textbooks provide equations to find a sufficient sample size, but they often do not include alpha and beta errors in the calculations. Moreover, because sample size is a function of five factors, it is usually found from tables rather than computed directly. For example, Table E.1 shows the required sample sizes for a plan with an alpha error of 10% and a beta error of 5%, and various acceptable and expected SDRs. The sample sizes indicated in Table E.1 were found from Equation E1.

$$n = \frac{\left\{ Z_a \sqrt{p(e)\left[1 - p(e)\right]} + Z_b \sqrt{p(a)\left[1 - p(a)\right]} \right\}^2}{\left[p(a) - p(e)\right]^2} \tag{E1}$$

Table E.1 Sample sizes for tests of controls.

	ADR (%)								
System rate (%)	2	3	4	5	6	7	8	9	10
0.1	203	123	87	66	53	44	38	33	29
0.5	*	220	139	99	76	62	51	44	38
1.0	*	*	225	148	107	83	67	56	48
1.5	*	*	*	216	147	109	86	70	58
2.0	*	*	*	*	203	143	109	86	71
2.5	*	*	*	*	*	190	138	106	85
3.0	*	*	*	*	*	*	177	132	103
3.5	*	*	*	*	*	*	229	165	126
4.0	*	*	*	*	*	*	*	208	154
4.5	*	*	*	*	*	*	*	*	190
5.0	*	*	*	*	*	*	*	*	239

Beta error = 5%; alpha error = 10%.
The positions marked with an asterisk are those in which the sample sizes are greater than 250, because these sizes were omitted as being impractical for discovery audits of controls.

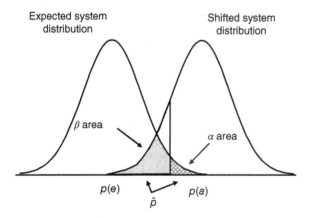

Figure E.2 Distributions of expected and shifted SDRs.

In Equation E1, the value of Z_a is 1.28, the number of standard deviations corresponding to a probability of $1 - \alpha$ in an upper-tail distribution, where $\alpha = 0.10$. The value of Z_b is 1.645, the number of standard deviations corresponding to a probability of $1 - \beta$ in a lower-tail distribution, where $\beta = 0.05$ (McClave & Dietrich, 1988). Then $p(e)$ is the estimated system rate and $p(a)$ is the ADR (see Figure E.2).

Table E.1 reinforces the fact that as the estimated SDR approaches the ADR, the required sample size gets larger in order to distinguish the two rates and maintain the prescribed risks. This is because a smaller interval between the deviation rates leaves less room for sampling error for given sampling risks of α and β.

Also, as the estimated SDR increases, quickly very large sample sizes are required, which may discourage the use of fixed rate sampling in controls from discovery.

The relationship of consumer risk, β, to the deviation rates is summarized in Table E.2. For example, if you want to decrease the β risk, you must increase the sample size. If you want to increase the ADR, the sample size can be decreased. If you increase the estimated SDR, or if the measured rate proves to be greater than the estimated rate, then you must increase the sample size and rerun the test. However, Table E.2 works both ways. If you then decrease the sample size, the beta risk increases because they are inversely related.

Figure E.2 may clarify how Equation E1 works. In this figure, the expected system distribution overlaps with some other undefined distribution. It could be a distribution centered at the ADR, and the SDR is nearer the ADR than we imagined. Or it could be that unknown to anyone, the SDR has shifted and the control is no longer effective.

Recall from Appendix D that an alpha error is a decision that the sample will indicate an effective control is ineffective. This will happen if the sample, $\hat{p}$, was drawn from the upper tail of the distribution centered at $p(e)$, for then it might be judged as not likely to derive from this distribution. This could occur in either tail, but for purposes of demonstration, we consider the system's upper tail. If $\hat{p}$ is drawn from the α area, it will be ≥ 1.282 standard deviations from the mean value (one-tail statistic).

Conversely, a beta error is a decision that the sample will indicate that an ineffective control is effective. Assume that $\hat{p}$ is drawn from the β area. This will appear to be in the $p(e)$ distribution; indeed, with a relatively high probability. Yet, it could be in the lower tail of the shifted distribution centered at $p(a)$, which represents an ineffective control.

Table E.2 The relationship of beta error and deviation rates.

Factor	Change	Sample size
Beta	Increase	Decrease
ADR	Increase	Decrease
SDR	Increase	Increase

E.1.6   Perform the Sampling Plan

The analyst begins taking samples. In the worst case, if a large number of deviations are found up front, it may be well to cancel the sampling plan as not cost-effective, and conclude that the control procedure is ineffective. If you are going to take 88 samples and you find 11 nonconformities in the first 20 or 30 tests, why continue the plan?

Generally, things will go according to plan if your pretest knowledge of the system is good. Assume that we expect a SDR of 1%, are willing to accept a deviation rate of 5%, and want to limit the beta error to 5%. In these conditions, Table E.1 gives a sample size of 148. If the measured deviation rate is approximately equal to the initial estimate, then this sample size is appropriate. If the measured deviation rate was 0.5%, then you find that the sample size should have been 99. You took a greater sample than was necessary but still conducted an effective test program. A pilot sample strategy would have been useful here, which would have indicated that fewer tests were going to be needed than originally supposed, and before you made them.

On the other hand, if you measure a deviation rate of 1.5%, then the sample size should have been 216, and you have another sample of size 68 to make. This increased requirement occurs because the interval between the deviation rates is too small to reliably discern the rate difference and you need larger samples to ensure the desired sampling risks. However, we can anticipate a problem here because a sample size of 216 approaches a size too large for many partial audits of controls.

E.1.7   Evaluate Sample Results

Initially, our estimate, $p(e)$, of the SDR came from historical data, pilot data, or knowledge of the process under test. After sampling, we update the estimate with the measured rate $\hat{p}$ determined from Equation E2, where x is the number of deviations found in the sample, n.

$$\hat{p} = \frac{x}{n},\tag{E2}$$

Assume that 2 deviations were found in a sample of 50. Then use the BETAINV application of EXCEL as shown in Appendix C to determine the highest likely deviation rate from the sample. Let $n = 50$ and $x = 2$. The sample deviation rate is thus 2/50, or 4%. Given this estimate, what is the highest probable deviation rate with a confidence of 0.95? In the BETAINV dialog box, enter the confidence, 0.95; enter the first shaping parameter, $x + 1 = 3$; and enter the second shaping parameter, $n - x = 48$. The result from BETAINV is the value 0.121, or 12.1%, as an upper deviation rate for this process.

The confidence level input to BETAINV is $1 - \beta$, so effectively the application uses the desired type II error in its results; in this case, a β-error of 5%. As an aside, this result agrees with the same upper deviation limit shown in tables by Apostolou & Alleman (1991c). You then compare 12.1% to the ADR to assess whether the control is conforming or nonconforming.

E.2 Stop-or-Go Sampling

The fixed-size attribute sampling plan is appropriate for audits of developing systems of operation because it can measure controls having relatively large deviation rates or modest expectations. Controls that are performing at a high level of performance will have a lower percentage of deviations and in such cases a sampling strategy is needed that can detect the relatively hard-to-find events. One of them is *stop-or-go sampling.*

Stop-or-go sampling is an attribute sampling strategy from the financial audit world and is used for financial controls whose expected deviation rate is very small. It is more efficient than a fixed-size attribute sampling plan because it uses small sampling numbers as long as the measured deviation rate is very low. The audited population is segmented into blocks, each block having a relatively small sample size. If the sample deviation rate remains small over a few blocks, the audit may stop. If the measured rate is higher than expected, the auditor has the option of increasing the sample size and continuing, or of stopping the audit of that control system.

Stop-or-go sampling may be appropriate to controls of operations because of the natural grouping of operational functions, say documentation, production, resources, and so on. If the control system has a low deviation rate and if attribute testing is performed on each group, then an average SDR can be easily determined. Hence, stop-or-go seems a natural choice as a sampling plan for forensic analysis of systems.

Only two factors are specified to conduct stop-or-go sampling: (i) confidence level and (ii) ADR. We discussed confidence level in Appendix C, so only the ADR will be demonstrated here.

E.2.1 Acceptable Deviation Rate

The ADR and its corollary, the acceptable performance level (APL), are system-level measures that are used frequently hereafter, so acronyms are convenient. Strictly speaking, the ADR is not a statistic. It is simply a declared value. Yet, it is derived from our knowledge of past system performance based on statistics, so it is not completely deterministic either. Stop-or-go sampling uses an ADR to create a sample size and to provide a standard of comparison for the measured SDR. In a binomial distribution, the probability that x events will occur in n trials is

$$P(x) = \binom{n}{x} p^x q^{n-x}, \tag{E3}$$

where p is the probability of a single event occurring, and $q = 1 - p$. From Equation E3, the probability of no events is

$$p(0) = q^n \tag{E4}$$

Since the events we are concerned with are deviations, then p is the probability of a deviation and q is the converse—a probability of a certain level of quality. Let the value p represent an ADR. Then q is a measure of an acceptable performance. For example, if $q = 0.99$, this implies that the SDR is 0.01. In order to determine the sample size needed for an audit, we specify an ADR as the maximum SDR that we will accept. Thus, we have an acceptable performance level (APL) defined as

$$\text{APL} = 1 - \text{ADR} \tag{E5}$$

(Careful readers will note that $\text{APL} = q$, and therefore $\text{ADR} = p$, but, rather than use the letters p and q of mathematical notation, I prefer to use the acronyms because they are descriptive terms of system performance.)

E.2.2 Sample Size

Operations systems that perform very well have, by definition, very low deviation rates. Samples with zero deviations can be expected. However, zero deviations should not be considered a norm. No system is perfect, but if there are just a few deviations, we want a sampling plan that will find at least one of them. Mills (1989b) meets this challenge with an alternate definition of confidence—the confidence that the sample size will detect at least one deviation, if any exists. This can be expressed as

$$C = 1 - p(0), \tag{E6}$$

where $p(0)$ is the probability of finding zero deviations in a sample of size n and a quality level of q. Although not readily apparent, Equation E.6 is entirely in keeping with the definition of confidence coefficient presented in Appendix C. To see this, refer back to Equation E4. For a given quality level, q, we can choose a sample size, n, such that $p(0)$ will have some arbitrarily low value. Values at $p(0)$ or lower become a rejection range.

It is critical to understand exactly what this confidence index, C, is—it is the confidence that the sampling plan can find a deviation, if such exists. It is *not* the confidence in the estimate of the SDR. At this point, we are only setting up the sampling plan. No samples have yet been taken.

Using the notations, C = confidence index and ADR = acceptable deviation rate, we can solve Equations E4 and E.6 for n:

$$n = \frac{\ln(1-C)}{\ln(1-\text{ADR})}. \tag{E7}$$

The terms C and ADR are used in Equation E7 in order to focus on the big picture and not bog down in p's and q's. Confidence indices and ADRs are system-level metrics. Knowing something of the system, we can specify C and ADR for any given audit, then easily determine the necessary sample size. Moreover, although you can solve Equation E7 with a spreadsheet, you can also solve it easily on a hand-held calculator, and it is a one-time per audit calculation.

As an example, suppose that for a system audit you choose a confidence level of 95% as your baseline, and from recent records, you determine the ADR is 0.04. Then using Equation E7, you will need a sample size of $n = 74$. Then if *no* deviations are discovered, the evaluation is thus: "There is a 0.95 probability that the SDR is less than or equal to 0.04." Table E.3 provides sample sizes for a range of confidence levels and ADRs.

Table E.3 Sample sizes for stop-or-go sampling (zero expected deviation rate).

	Confidence levels		
	90%	95%	99%
ADR (%)	Sample sizes		
10	22	29	44
9	25	32	49
8	28	36	55
7	32	41	64
6	37	48	74
5	45	58	90
4	57	74	113
3	76	98	151
2	114	148	228
1	*	*	*

An asterisk means that the sample size too large to be useful for discovery audits of controls.

Comparing Tables E.1 and E.3 shows that good systems require smaller sample sizes and hence less auditing than developing systems, and this satisfies our moral sense. There are rewards for maintaining good systems, and this is one of them. However, even good systems sometimes stray, and the proof is in the evaluation, to paraphrase an old saying. Stop-or-go sampling plans respond appropriately if a system is beginning to drift in effectiveness, as demonstrated in the next section.

Table E.3 indicates that even with stop-or-go sampling, the nearer the expected deviation rate is to the ADR, the larger must be the sample in order to distinguish the difference between them.

E.2.3 Evaluation

In the event that a deviation is found where none were expected, financial auditing has a set of decision rules on whether to stop or go, and if to continue, how large an additional sample must be. However, operations analysis takes a different approach. As in fixed-rate sampling, we are not so much concerned with estimating the true SDR from a single sample as we are in determining its probable upper limit—the worst-case scenario, so to speak. Worse case scenarios help to frame litigation strategy.

Again, you repeat the spreadsheet process, using BETAINV and choosing a $\beta = 0.05$. Suppose that you find one deviation in a sample of 67. Then the upper deviation limit is 6.9%. If two deviations were found, the upper deviation limit is 9.1%. The result, of course, would be compared to the ADR for a decision on control conformity.

E.3 One Hundred Percent Inspection

One hundred percent inspection means that there is no sampling at all—every item of the lot is inspected. Hence, there can be no sampling errors. However, it can be easily shown that 100% inspection of a control system is less than 100% effective.

Assume that in 500 previous tests, an auditor failed to detect one nonconformity. That results in an inspection failure rate of 0.2%, quite low; indeed, it would qualify as a rare event. This inspection is 99.8% effective. However, given this past rate, the analyst makes an additional 100 tests in a new test of the system. Then the total effectiveness of the inspection is $(0.998)^{100} = 0.82$. That means an 18% beta error, far worse than any formal sampling plan. This beta error compares closely with the estimate of 80% effectiveness of 100% inspection plans cited by Joseph Juran (1935).

Apart from its inherent inspection failure rate, 100% inspection is strongly criticized by efficient experts as unnecessarily costly. Yet, there are times when

100% inspection is necessary, for example if it is required by the customer or if a process has a history of poor performance. And even on a cost basis, 100% inspection can be justified by simply comparing the cost of no inspection to the cost of inspection (Grant and Leavenworth, 1988; Schorn, 2012; Taguchi et al., 1989). Cost analysis is beyond the scope of this book, but the reader might wish to pursue the subject with the given references.

E.4 Application: An Attribute Sampling Plan

Wild Rover, Inc. is an ISO 9000-certified company whose contract performance is in litigation. In particular, its documentation control procedures are being questioned. The strategy of the forensic systems team is depicted in Table E.4, showing the documents, owner-functions, and estimated document populations to be examined. In adjacent columns, the relative percentage of the subgroup population is also shown, with the associated number of documents to be drawn at random from that subgroup population. The following considerations were included in the audit plan:

- A control test confidence level of 95% was required.
- A deviation rate maximum of 5% would be the tolerable rate.
- The expected deviation rate would be low, about 1%.

Column 1 displays the company functions to be audited. At *Wild Rover*, all functions are performed in a departmental structure. Column 2 displays the document type that will be audited and to which appropriate questions are addressed in the checklist. Column 3 lists the population of documents from

Table E.4 Sampling strategy of the *Wild Rover* forensic team.

Function	Document type	Document population	Percent of population	Sample size
General manager	Management reviews	15	0.416	3
All departments	Quality manual	80	2.22	10
Acquisitions	Purchase orders	250	6.94	18
Production	Job orders	1500	41.6	50
	Calibration records	60	1.67	8
Customer service	Delivery orders	1500	41.6	50
Personnel	Training records	200	5.55	15
Total		**3605**	**100**	**154**

which samples will be audited. This population is defined not only by type of document but also by period of time. The documents will be examined for the previous 12-month period, the time period in litigation. For example, it is known that there were 15 management reviews in this period, all of them recorded. Three of them will be examined by the forensic team. The total number of documents is 3605, which is considered a "large" population for statistical purposes. Indeed, the control population may be much greater than this if each document has several controls to it, for example, authorization, revision date, and clarity. The focus of the investigation of controls will depend to a large extent on the nature of the complaint.

Column 4 shows the percentage of each function population to the overall document population. Since the acceptable and expected deviation rates have been declared, 5 and 1% respectively, the required sample size can be found from Table E.1. The required sample size for this audit is 148.

Accordingly, the percentage of 148 to be sampled for each subgroup is determined and rounded off to the next highest integer. Notice that the sample size actually taken is 154. This is due to rounding up at each subgroup, to ensure that the individual function meets its own sample size criteria.

After finishing the inquiry phase, the forensic team sums up the number of deviations found in *each department.* Assume that it is four deviations total, over the sample of 154, arriving at an estimate of 2.6% as the deviation rate of documentation control. However, a spreadsheet evaluation yields an upper deviation limit of 5.8%, just over the ADR. Thus, the forensic team can be 95% confident that the *Wild Rover* documentation control system exceeds the acceptable level, although given the small margin over the ADR, the decision will almost certainly be challenged.

Note that the test is of the effectiveness of the total documentation control system and not of the documentation control of any one function. Some functions will have better rates, some worse. The forensic team will next analyze the performance of each department in turn, to focus on root cause.

References

Apostolou, B. and Alleman, F. (1991a). *Internal Audit Sampling.* Altamonte Springs, FL: The Institute of Internal Auditors, p. 45.

Apostolou and Alleman (1991b), p. 47.

Apostolou and Alleman (1991c), p. 115.

Grant, E. L. and Leavenworth, R. S. (1988). *Statistical Quality Control.* New York: McGraw-Hill, pp. 619–621.

Hogg, R. V. and Tanis, E. A. (1983). *Probability and Statistical Inference.* New York: McMillan, p. 377.

Juran, J. M. (1935). "Inspector's Errors in Quality Control." *Mechanical Engineering*, vol. 57, p. 643.

McClave, J. T. and Dietrich F. H. (1988). *Statistics*. San Francisco, CA: Dellen, p.332.

Mills, C. A. (1989a). *The Quality Audit*. New York: McGraw-Hill, p. 171.

Mills (1989b), p. 174.

Schorn, T.J. (2012). "Management Decision Making and the Cost/Benefit of Multiple 100% Inspections." *AFS Transactions*, Paper 12-049, American Foundry Society, Schaumberg, I, pp. 33–46.

Taguchi, G., Elsayed A., and Hsiang, T. (1989). *Quality Engineering in Production Systems*. New York: McGraw-Hill, pp. 20–22.

Appendix F

Nonstatistical Sampling Plans

A sampling plan is nonstatistical when it fails to meet at least one of the criteria of a statistical sampling plan (Apostolou & Alleman, 1991a). Yet, there are similarities between the two types of plans. Both require judgment on the part of the analyst. Audit procedures will be the same for either type of plan, and both are permissible by the standards of the Institute of Internal Auditors (IIA, 2013). The major difference between them is that sampling risks cannot be measured in a nonstatistical sampling plan.

There are circumstances in which nonstatistical sampling is the more appropriate of the two. Nonstatistical sampling may be used when results are needed to confirm a condition, or when judgment is more important than random sampling, such as in reviewing records, searching for unusual relationships, or interviewing client personnel (Guy, 1981a). Nonstatistical sampling is also correct when there is no sampling—the entire lot is tested. This seems to be an oxymoron; a sampling procedure for a non-sampled population. Yet, sampling plans bring a formality and direction necessary to any analysis. Consider that even if you intend to test the entire lot, it may be that some of the evidence is trivial or irrelevant and will be abandoned, in which case the balance becomes a sample.

Forensic Systems Engineering: Evaluating Operations by Discovery,
First Edition. William A. Stimson.
© 2018 John Wiley & Sons, Inc. Published 2018 by John Wiley & Sons, Inc.

The analyst should understand that a nonstatistical plan does not mean that statistical methods are not used. On the contrary, they are used. Rather, it means that sampling error cannot be measured in a nonstatistical plan, so conclusions based on such statistics are easily challenged. But statistical methods, sampling error or not, *can* be used to provide direction to corroborating evidence.

Although nonstatistical methods cannot measure the sampling risks, it is possible to derive important conclusions from the results of a sample, using information derived from the control parameters. This is possible if the test is a Bernouilli trial—a random experiment in which there are only two mutually exclusive outcomes and in which the probability of a given event is constant. Within the context of this book, the mutually exclusive events are conformity and nonconformity. The constant probability of concern is the rejection or deviation rate, that is, a control that is nonconforming to requirements (Grant & Leavenworth, 1988). The rate has a binomial distribution about which limiting probabilities can be estimated, such as upper deviation rates, which is another reason for using statistical rules whenever possible in a nonstatistical sampling plan.

F.1 Sampling Format

The sampling format for a nonstatistical plan is the same as for a statistical plan and consists of framing the dimensions of the sampling: its objectives, scope, and basis. The primary interest here is how statistical procedures might fit into a nonstatistical analysis, but a holistic approach to the question increases the confidence level of a nonstatistical plan. Qualitative assessments may be used in lieu of the quantitative approach used in statistics, with necessary precautions and supporting evidence.

The format will focus on those characteristics of internal control deemed essential to acceptable performance. The nonstatistical sampling plan most relevant to this book is an attribute plan and the same seven-step strategy suggested in Appendix E for statistical sampling is used in a nonstatistical audit:

- Determine the objectives of the test
- Define the attribute and deviation conditions
- Define the population
- Determine the method of sample selection
- Determine the sample size
- Perform audit procedures on the sample
- Evaluate the results and state conclusions

F.1.1 Frame of the Sampling Plan

A sampling plan is framed by its purpose, scope, and basis. The purpose of the plan is, whether statistical or nonstatistical, within the context of this book, to assess the performance of internal controls.

The scope of the analysis is the set of functions or processes to be assessed. In the case of assessing the operations of a corporation, an approach can be general, in which all activities are analyzed, or specific, in which the focus is on a subset of the activities. The subset may be selected because of an inquiry into an area of special concern, or perhaps because of an identified set of consistent problems. But even if the analysis is general, the activities to be assessed must be specified if the analysis is to be bounded.

The basis of the analysis is the set of applicable standards, rules, or regulations from which the assessment is evaluated. These requirements will be mapped into each activity according to their applicability. This is called a vertical audit. If you are going to analyze a purchasing function of the company, then requirements applicable to the purchasing function are appropriate and the requirements applicable to, say, design reviews are not.

F.1.2 Attribute and Deviation Conditions

The attribute and deviation conditions are the same as those for statistical sampling. An internal control is deemed either conforming or nonconforming to requirements. The conditions of interest are deviation rates and are estimated as described in Chapter 14: in terms of major, moderate, or minor risks to effective control. This is a subjective decision tempered by the period of time in which the nonconformity exists. A single event may be simply regarded as an outlier, but risk increases with time and an extended period of nonconformity is always a major risk.

F.1.3 The Population

The population will be a system or subsystem of internal controls of the operations under study. The sample population will probably, but not necessarily, be a subset of the total. It is possible that the entire lot will be tested, in which case the study is no longer a sample, but statistical sampling procedures are nevertheless continued in order to enhance the credibility and validity of the results.

The sampling procedures are where the choice of nonstatistical and semi-statistical activity comes into play and are the same for both statistical and nonstatistical plans. Even though you do not use probability laws in a nonstatistical sampling plan, using statistical procedures enhances objectivity. In keeping with statistical discipline where appropriate, records for inquiry should be done randomly even if the sample size is too small to make assumptions about the distribution of the greater population.

A population is a set of objects or persons possessing the characteristic(s) that are being studied. A representative sample is a subset of the population, each member of which possesses the characteristics that represent the entire population. A simple analogy is that of a sample representing a population of oranges. The sample should not contain apples. However, this analogy is too simple; differences among members can be quite subtle and is greatly influenced by the objectives and characteristics of the study. For example, suppose the effect of a certain medication is under study. Depending upon the study objective and the nature of the effect, samples taken from a population of those using the medication may not be representative if some members of the sample were young adults and some were senior citizens.

Sometimes a representative sample is called a random sample, but the two are not the same. A random sample is one taken from a lot in which each member of the lot has an equal probability of being selected. Thus, a sample is selected from a lot in such way that it is homogeneous, representative, and random.

F.1.4 Nonstatistical Sample Selection

The selection of samples is made either by haphazard, blocking, judgment, or stratified techniques. Haphazard selection is a selection made without bias, and this is best done through randomness if possible. You must be satisfied that the sample is representative of the population. Defined probability concepts are not used, nor is statistical inference.

Block selection means that you select a homogeneous subgroup, say all the job orders for a given day or week. To ensure fair representation, many blocks should be selected. Randomness can be introduced by randomly selecting the days or weeks for which you will examine the job orders. For example, suppose you want to examine job orders over the last 30 days. Then using block selection, you choose all the job orders for several randomly selected sets of, say, 4 or 5 days, depending upon your time available.

Judgment means that samples are taken from areas that you believe have high control risk, either due to the nature of the process or due to past audit results. Shipyards provide a good example. A typical US Navy ship repair contract will contain over 5000 work items, but only about one-tenth of those work items will be on the critical path. These are called controlling work items, and their status provides a good estimate of repair progress. Once you have selected a set of controlling work items or similar set of high-risk items for examination, then you should randomly select from that pool. This is where the notion of materiality comes into play. It is a departure from randomness but may be needed for effective selection.

In regard to materiality, there are three definitions that are pertinent to the sampling of evidence: (i) Materiality is a measure of the effect that an item of information will have on the accuracy or validity of a judgment; (ii) Materiality is the threshold of importance of a given process on the performance of operations; (iii) "Information is material if its omission or misstatement could influence the economic decisions of users taken on the basis of financial statements" (IASB, 2001).

The first definition refers to a judgmental error on the part of an analyst vis-à-vis the conformity of a control. The second refers to the importance of a given control relative to the operations in question.

The third definition refers to the financial importance of the control. Investors purchase shares in a company based in part on its price/earnings ratio. Thus, materiality is tied to that value of a company's estimated worth that could influence investors. But materiality also reflects corporate operations because financial statements are about the costs of doing business. They include all the costs associated with operations, which itself is a function of the quality of its performance. Thus, all three definitions come together in the judgment of the analyst of a system of controls of operations.

F.1.5   Sample Size

In statistical sampling, the choice of sample size was influenced by five parameters: the expected system deviation rate (SDR); the acceptable deviation rate (ADR), the desired Alpha and Beta error probabilities, and the desired confidence level. An alpha error (type I) occurs when the analyst assesses a control as ineffective, when in reality the control is effective, with little nonconforming output. A beta error (type II) occurs with an assessment by the analyst of a control as effective, when in reality the control is ineffective.

Thus, Alpha and Beta errors are made in the evaluation and analyses of sample results and will happen in nonstatistical audits as well. However, these errors cannot be measured in nonstatistical sampling because not all of the conditions required of statistical inference are present. Nevertheless, you need to know, to a "reasonable" degree, whether a control is effective or not.

A nonstatistical plan, properly conducted, will yield a valid proportion of deviations to sample size and you are free to preselect an ADR. It is the interval between these two rates that cannot be accurately measured, but you can still make a good estimate of control effectiveness. There is good agreement among audit experts on the assignment of numbers to a qualitative assessment of a control. Table F.1 compares the value system of several experts, such as Kelly (1986) and Guy (1981b).

Table F.1 is read in this way. Using qualitative judgment, you select a desired confidence level, say "substantial." Then the table provides the ADR that can reflect this level. Then armed with this rate and an expected SDR, you can select sample size from Table E.1.

The levels of the ADRs of Table F.1 may seem unusually large to people in operations, but they must be understood in context. They apply to the deviation rate of *controls*, and not to the deviation rate of key characteristics of a product or service. Table E.1 does not provide sample sizes for rates greater than 10%, but they can be determined from Equation E.1 if occasion demands.

Table F.1 Selecting audit confidence level as a function of the ADR.

	Acceptable deviation rate	
Assessment	Kelly	Guy
Substantial confidence	2–7%	≤5%
Moderate confidence	6–12%	≤10%
Little confidence	11–20%	≤20%
No confidence	Omit test	Omit test

However, Guy (1981c) advises against using ADRs greater than 10% if the purpose of the inquiry is about the reliance of internal controls, as opposed to some other of their characteristics. As reliance of a control is a major issue in systems operation, then 10% can be considered an upper limit for ADRs for purposes of forensic systems examination.

As in statistical sampling, in a nonstatistical plan you use knowledge of past system performance to estimate the control deviation rate in order to arrive at a sample size. Then having chosen the sample size and conducted the test, you evaluate the control based on the difference between the sample and the ADRs.

F.1.6 The Effect of Sample Size on Beta Error

Many analysts consider Beta error as the more grave sampling error, as Alpha error tends to be self-correcting. If an analyst makes an Alpha error, the decision will be challenged because the analyst is challenging a control that the performer believes to be good. The performer will insist on additional testing with larger sample size. However, a Beta error will not be challenged; the analyst reports good news to the performer, although the good news is false. In addition, additional testing increases cost. Apostolou (1991b) classifies an Alpha Error as an efficiency indicator; a Beta error as an effectiveness indicator.

Nonstatistical sampling plans do not measure Beta error, but you know that it exists. You must make some effort to reduce it, and sample size has much to do with it. Sample size is affected by the error rate as shown in Table E.2, in which changes in sample size affect key system parameters and conversely.

For example, suppose that you want to choose a smaller sample size. By doing so, you increase the chance of Beta error—the risk of assessing a control as effective when, in fact, it is not, because sample size and Beta error are inversely related. As another example, if you want to increase the ADR, then you can decrease the sample size. But in doing so, you increase the chance of Beta error. The reason should be clear. A larger ADR offers a less effective control, or a less accurate sample of a good control. But if you want to tighten the system assessment by decreasing the ADR, then a larger sample population is needed, again to err on the fail-safe side.

As a final example, if you expect a low SDR, then you can use a lower sample size, understanding that the lesser accuracy is unimportant for a very good system. But there, again, the chance of Beta error increases. On the other hand, if you expect a high system error rate, possibly above the acceptance rate, then you need a larger sample size to be sure of the margins. These considerations require judgment, experience, and knowledge of the system, but they enhance the validity of a nonstatistical plan.

F.1.7 Evaluating Sample Results

Because nonstatistical sampling does not provide a reliable estimate of sampling risk, the analyst must make a judgment of whether the difference between the ADR and the measured deviation rate is an adequate allowance for sampling error. For example, suppose the analyst expects a SDR of about 1% and will accept a deviation rate (ADR) of 7%.

From Table E.1, a sample of 83 is taken. Suppose further that in this sample, two errors are found. Then the measured deviation rate is $2/83 = 2.4\%$. The analyst must then decide whether the 4.6% difference between the measured rate and the ADR is sufficient to cover sampling errors.

Here is where the more you know about the process the better will be your judgment. A sample of 2/83 is about as likely to derive from a distribution centered at 5% as from a distribution centered at 1%. In this case, the measured deviation rate (2.4%) was more than double the expected rate (1%), so a test of the 95% confidence level is called for. A BETAINV test of 2/83 shows the result to have a 95% confidence that the true SDR is less than 7.4% (Apostolou, 1991c). As this limit is just over the ADR, the control should be deemed effective in order to avoid an unwinnable challenge in litigation.

Thus, the bounds and confidence levels of a test of controls are critical and must be justified, with generous allowance made for uncertainty. For example, if the confidence level were set at 90%, the upper limit of the SDR at 2/83 would have been 6.8%, thus below the ADR. If the expected SDR were 1.5% instead of 1%, the upper limit of the SDR at 2/109 would have been 5.7%, well below the ADR.

F.2   Nonstatistical Estimations

Sampling conclusions are always estimations, irrespective of the type of sampling plan that is used. Given that caution, statistical sampling plans offer a means to estimate the errors in comparing the sample to the population. Nonstatistical sampling plans cannot estimate measurement errors and require extraordinary understanding of the processes being monitored and larger margins between ADRs and estimated deviation rates.

Nevertheless, numerical results from nonstatistical sampling are useful when framed to answer the right questions: (i) "What is the highest SDR that is likely to yield the sample?" (ii) "What is the smallest Beta risk that is likely, given the sample?" Still, forensic systems engineering is about preparing engineering data for trial. All the conclusions based on the analysis of evidence can be subject to challenge and rebuttal. Therefore, when using a nonstatistical sampling plan the benefit of doubt in close margins of deviation rates goes to the defense. Chapter 17 addresses this dilemma and recommends a forensic focus on systemic failure when statistical conclusions derived from the evidence contain controversial margins in probable sampling errors.

References

Apostolou, B. and Alleman, F. (1991a). *Internal Audit Sampling*. Altamonte Springs, FL: The Institute of Internal Auditors, p. 11.

Apostolou (1991b), p. 9.

Apostolou (1991c), p. 60.

Grant, E. L. and Leavenworth, R. S. (1988). *Statistical Quality Control*. New York: McGraw-Hill, p. 201.

Guy, D. M. (1981a). *Introduction to Statistical Sampling in Auditing*. New York: John Wiley & Sons, Inc., p. 8.

Guy (1981b), p. 140.

Guy (1981c), p. 46.

Institute of Internal Auditors (2013). Practice advisory 2320-3: Audit sampling, p. 2. https://www.iia.nl/SiteFiles/PA_2320-3%20(1).pdf. Accessed September 13, 2017.

International Accounting Standard Board (2001). Framework for the Preparation and Presentation of Financial Statements, p. 83. http://mca.gov.in/XBRL/pdf/framework_fin_statements.pdf. Accessed September 13, 2017.

Kelly, J. W. (1986). *How to Use Statistical Sampling in Your Audit Practice*. New York: Matthew Bender.

Index

All references are with respect to page numbers. If the reference is a figure, the page number will be in italics. If the reference is made to a table, the page number will be in bold print.

Three entries are giants in American industry: Henry Ford, Steve Jobs, and Dave Packard. They are so listed in the index.

The standards: ANSI/ISO/ASQ 9000 (2005), 9001 (2015), and 9004 (2009) are fundamental to this book and appear frequently in many chapters. Therefore, they are listed in this index only when they are used as a reference in the narrative.

Forensic Systems Engineering: Evaluating Operations by Discovery,
First Edition. William A. Stimson.
© 2018 John Wiley & Sons, Inc. Published 2018 by John Wiley & Sons, Inc.